ALLE ZEIT WACH
SJ
1842

H. Kuckuck · G. Kobabe · G. Wenzel

Fundamentals of Plant Breeding

With the cooperation of
D. Böringer · W. Hondelmann · V. Stoy · T. Tatlioglu

With 65 Figures

Springer-Verlag
Berlin Heidelberg New York London Paris
Tokyo Hong Kong Barcelona Budapest

Professor Dr. Dr. h.c. HERMANN KUCKUCK
Reihermoorweg 5
W-3006 Burgwedel 4, FRG

Professor Dr. GERD KOBABE
Universität Göttingen
Institut für Pflanzenbau
und Pflanzenzüchtung
von-Siebold-Straße 8
W-3400 Göttingen, FRG

Professor Dr. GERHARD WENZEL
Institut für Resistenzgenetik
Biologische Bundesanstalt für
Land- und Forstwirtschaft
W-8059 Bockhorn, FRG

Contributors:

Dr. D. BÖRINGER
Presidente of the Federal Plant
Varieties Office, Germany
Osterfelddamm 80
W-3000 Hannover 61, FRG

Professor Dr. W. HONDELMANN
Schwetzingenstraße 14
W-3300 Braunschweig, FRG

Priv. Doz. Dr. V. STOY
Svalöf AB
S-26800 Svalöv, Sweden

Professor Dr. T. TATLIOGLU
Institut für Angewandte Genetik
Herrenhäuser Straße 2
W-3000 Hannover 21, FRG

Cover illustration: Combination breeding by application of the bulk population method (s. Fig. 10, page 29).

Title of the original German edition:
Hermann Kuckuck/Gerd Kobabe/Gerhard Wenzel
Grundzüge der Pflanzenzüchtung, 5. Auflage
© Verlag Walter de Gruyter & Co., Berlin 1985

ISBN-13: 978-3-642-75394-7 e-ISBN-13: 978-3-642-75392-3
DOI: 10.1007/978-3-642-75392-3

Library of Congress Cataloging-in-Publication Data. Kuckuck, Hermann, 1903– [Grundzüge der Pflanzenzüchtung. English] Fundamentals of plant breeding / H. Kuckuck, G. Kobabe, G. Wenzel; with the scientific cooperation of D. Böringer ... [et al.] p. cm. Translation of: Grundzüge der Pflanzenzüchtung. Includes bibliographical references and index. ISBN 0-387-52109-7 1. Plant breeding. I. Kobabe, Gerd. II. Wenzel, Gerhard. III. Title. SB123.K813 1991 631.5'3–dc20. 91-7971

Softcover reprint of the hardcover 1st edition 1991

31/3145-543210 – Printed on acid-free paper

Preface

All meat has been grass. This statement of the old Greeks is still valid and thus plants contribute to the needs of a growing world population. Their other key role for human life is reflected by their photosynthetic activity. Intensive agriculture reduces the carbon dioxide content of the atmosphere. The improvements of cultivated plants towards stable yields and desired quality characteristics are the subjects of plant breeding. Here, an introductory picture of the present state of this topic in research and application is given.

The senior author, H. Kuckuck, wrote the first edition in 1939 as a small German booklet, hence the nickname little cuckoo (Kuckuck = Engl. cuckoo). During the last 6 years (the fifth German edition appeared in 1985) again numerous results of basic research have found practical application, making extensive additions necessary. Further, it seemed timely to change to the international language of science: English. For help in translation we are grateful to Silke Kluth and Kathy Seaman.

Since agricultural sciences – like all others – have become more specialized, we were happy to welcome D. Böringer (Sect. 5.4), W. Hondelmann (Sects. 4.4, 4.5, and 6), V. Stoy (Sect. 4.2), and T. Tatlioglu (Sects. 2.3.1.4 and 3.5) as contributors to this edition.

Finally, we would like to thank Springer-Verlag for all their efforts in helping to transform a small German booklet into a book available to a wider audience.

June 1991 H. KUCKUCK, G. KOBABE, G. WENZEL

Contents

1 Introduction

Plant cultivation is not only the basis for the nutrition of man and animals, but it also provides industry with chemicals and other nonfood, crude products. It is obvious that farmers constantly try to increase the yield and quality of their products by applying more effective techniques. This can be done by varying the environmental conditions and, depending upon the requirements of the market, by optimal crop management.

In addition, the phenotypes of plants are determined by genetic factors. Traits are inherited from parents by their progenies and, by applying certain procedures, the transfer of genetic factors can be influenced. After the rediscovery of Mendel's laws, the knowledge of heredity increased rapidly, and genetics became the fundamental science for the activities of those who try to change the genetic constitution of plants and, thus, to breed new and better varieties for farmers. Therefore, plant breeding can be interpreted as applied genetics.

One prerequisite of successful plant breeding is genetic diversity. The composition of a population consisting of plants with different genetic constitutions can be changed by selecting distinct phenotypes. Selection was the earliest method of plant breeding and it was practiced for centuries before man learned to understand the heredity of plants. After the discovery of sex in plants, hybridization became an additional breeding technique. Later, plant breeders found that after crossing distinct genotypes, the vigor of the F_1 generation was highest, higher than that of the parents, higher than that of the basic material, and higher than that of the subsequent generations. Many attempts were made to exploit this hybrid vigor, which was due to heterosis. When genetic diversity is not satisfactory, genetic variability can be broadened by the induction of mutations.

In recent years, remarkable progress has been made in molecular genetics and in tissue culture. The results of this field of biology provide valuable contributions to plant breeding. After the cultivation of small pieces of the plant on special media, functional plants can be regenerated. If anthers or microspores are used, the production of haploid plants is possible and, after polyploidization, true homozygous plants can be established. On the other hand, plants can be separated into single cells and, after dissolution of the cell wall, the protoplasts are able to fuse, which means that different species can be combined on the somatic level. Furthermore, single genes can be identified and transferred from one genotype to another.

In the historical sequence mentioned above, plant breeding methods were developed. But new methods did not replace the older ones. New methods rather constitute an extension of the tools that can be used in breeding programs.

Since plant breeding cannot be carried out by the application of genetics alone, other basic plant sciences such as plant cultivation, plant physiology, and plant pathology support effective plant breeding work.

The purpose of plant breeders' work is the steady improvement of varieties. The most important fundamental methods are described in the following chapters.

2 Basic Breeding Methods

Based on the genetic methods that are being applied in the utilization and creation of genetic variation and in the selection of stable variants, breeding methods can be classified into three groups: selection breeding, combination breeding, and hybrid breeding. As is the case in all classification schemes, there are fluid transitions and similarities between the individual groups of breeding methods. Selection not only takes place in the group termed selection breeding, but also in the other groups. They mainly differ from one another in the methods with which the varying original material for breeding is produced. In breeding by combination, new variation is generated by crossing genetically different parents. In hybrid breeding, inbred lines of different genetic constitution are obtained by selfing, pair crossings, and other methods, and from these, those lines resulting in the best combinations on the basis of test crosses are selected.

In breeding by combination as well as in hybrid breeding, a cross takes place. However, while in combination breeding the desired product must be bred to constancy, in hybrid breeding a heterozygous product is marketed as a variety that is continuously reproduced by the breeder for the grower. The subsequent generation seed is not constant; further cultivation is therefore not possible.

The three groups of breeding methods have been listed here in the chronological order of their development. While breeding by selection originally was developed purely empirically without any scientific preparatory work, and only later was effectively altered by applying genetic knowledge, combination breeding very quickly developed on a scientific basis after the rediscovery of Mendel's laws at the beginning of this century. Soon after this, intensive development of hybrid breeding occurred, especially with regard to its various modifications.

In addition to the above-mentioned classifications, breeding methods can also be classified according to other criteria. Thus, Schnell (1977), for example, has classified breeding methods according to the manner of propagation of the varieties:

- clone varieties (vegetative propagation);
- line varieties (propagation by self-fertilization);
- panmictic varieties (propagation by open pollination);
- hybrid varieties (continuous reproduction by controlled crosses).

When selecting a breeding method, the breeder is not guided by the criterion of its newness but only by its suitability and profitability. The breeding method to be applied is mainly determined by the aim of breeding, by the available original material, by the general state of breeding of the country in question, by the available technical and scientific equipment and, finally, by the proficiency of the breeder himself.

Two important steps in the course of breeding shall be briefly pointed out: in each breeding method, variability and selection takes place. The more reliable the available evaluation methods are for the desired attributes of the variants to be tested and the easier they can be handled, the greater the number of investigated variants can be and the greater also are the prospects of success.

With the establishment of a valuable seed stock, the work of the breeder is still not finished. In addition, knowledge of the requirements of the new variety concerning climate, soil, fertilizer, etc. is necessary, as well as good maintenance breeding and effective organization regarding the multiplication and distribution of the needed seed.

2.1 Selection Breeding

In order to succeed in selection breeding, the breeder must have original material at his disposal from which he can select those types most appropriate to his breeding aim. Above all, old local varieties, so-called landraces or primitive cultivars, are possible sources for selection.

Landraces are population mixtures containing a great number of different hereditary types which, due to their genotypic diversity, are especially well adapted to the annual changes in the environmental conditions of their habitat. Compared to modern cultivars, they have only average, but reliable yields. Today, in countries with more or less highly developed agriculture, these landraces have largely been superseded in cultivation by highly selected varieties; they can still be found predominantly in areas with primitive agriculture. Since landraces include many hereditary variants that have already passed through the sieve of natural selection for years. The qualitative value of landraces for selection is mainly determined by the geographical location from which they originate. Thus, a breeder who intends to breed a particularly winter hardy wheat will prefer land varieties from Canada, Sweden, Finland, or northern Russia for selection, since these quite probably include types with highly heritable hardiness.

With respect to heredity, the landraces differ depending on whether the plant species in question is self- or cross-pollinated. The landraces of self-fertilizing crops consist of a more or less large number of mainly homozygous types, the individual progenies of which are termed "pure lines." Only randomly occurring cross-pollinations between pure lines result in heterozygous plants, segregating new homozygous types in the following generations due to self-pollination, which can gradually supplant the heterozygous ones. On the other hand, landraces of outbreeders consist mainly of heterozygous types, since constant homozygous genotypes only rarely occur as a result of continuous cross-pollination. In outbreeding crops a pollen mixture of a different heritable nature is always involved in pollination. Therefore, the landraces of outbreeders do not consist of clearly distinguishable types as in self-fertilizing crops, but often display a series of types that smoothly merge into one another. Clear boundaries between self- and cross-pollinated crops cannot always be made. The portion of outbreeding in normally cleistogamous plants (self-fertilization within the closed flower), such as wheat, tomato, and others, depends on the genotype of the variety as well as on the respective environmental conditions. Thus, varieties in continental areas tend to cross-pollinate more vigorously than in Atlantic climates. In certain species, the extent of cross-pollination is determined by the presence of insects that transfer the pollen grains.

Within a species, types with strong morphological differentiation often occur. If they also differ from one another physiologically and thus prove to be especially well adapted to local climates, they are termed ecotypes or ecospecies, depending on the extent of their taxonomical differentiation. The individuals of an ecotype are only uniform in regard to the characters that provide them with special adaptation to certain environmental conditions; in all other characters they can vary.

Ecotypes can be found in perennial clover, lucerne, and grass species, which are important forage crops. In grass and clover species, varieties derive from direct selection of ecotypes from wild-growing plants. Therefore, in these species, there are not always clear boundaries between cultivated and wild species. The extent to which varieties with high forage yield still have characters of a wild plant becomes apparent during seed multiplication. Due to the frequently occurring disintegration of ripe ears or panicles or because of easy shattering, harvesting the seeds is difficult.

2.1.1 Mass Selection

In mass selection, the easiest method is to select and multiply together those types from a mixture of phenotypes corresponding to the breeding aim. This process is called *positive mass selection*. If only all undesired offtypes are rogued in a field crop and the rest are then propagated further, it is called *negative mass selection*. Today, this is no longer regarded as an adequate breeding method for developing new varieties; it is only used in the multiplication of established varieties, i.e., in field crops that are meant for seed production and that must be free of mechanical defects, diseased plants, and casual hybrids.

2.1.1.1 Positive and Negative Mass Selection

Positive mass selection is a selection according to phenotype, i.e., according to outer appearance. Since plants of the same phenotype very often do not have the same hereditary nature (ideotype, genotype), mass selection is not fully effective in all cases.

This method is applied in such a way that all plants corresponding to the ideal type of selected from a field crop and bulked as a population, and this process of repeated selection within the improved population is continued for several years. The progeny of each mass selection in cross-pollinated species should be cultivated in spatial isolation from crops of the same species, in order to prevent uncontrolled pollination and, consequently, the risk of genetic deterioration of the selection. In general, distances of 300 m for insect-pollinated species and 500–1000 m for wind-pollinated species should be sufficient.

The positive mass selection method can quite quickly result in notable success in the improvement of the cultivated landraces, e.g., in developing countries. It is especially recommended if the technical prerequisites for employment of other, more complicated methods are not yet available. Nevertheless, the application of positive mass selection is still justified, for example, in the improvement, with respect to uniformity, of cross-pollinated vegetable species such as carrots, radishes, beetroots, etc. Mass selection carried out with high standards and on a large scale may contribute more to attaining this goal than a progressive method carried out with insufficient means and on a limited scale, e.g., selection of individual plants with subsequent progeny trials (pedigree method).

The effectiveness of mass selection depends upon the type of genetic control of the characters to be selected, i.e., whether or not the character is strongly influenced by environmental conditions. Furthermore, the success of the selection is influenced by the reproductive biology of the species in question, whether or not it is self- or cross-pollinated and if, after cross-pollination, the characters to be selected are discernible before or after flowering. In qualitative characters with alternative inheritance, which are mostly monogenic and easily identifiable, mass selection for *recessive* characters within a segregating progeny quickly leads to success, whereas selection for *dominant* characters is only partially successful. When selection is continued, the proportion of undesired types decreases. This is more quickly accomplished in cross-pollinated species, where selection is possible *before flowering*, than in outbreeding species, where selection of the character in question can only take place *after flowering* (Table 1).

In *quantitative* characters, which are polygenic and therefore transitional, success is determined by the number of genes controlling the characters and by the degree of the modifying influence of the environment. In order to be fully effective, mass selection must also be carried out on a sufficient scale from as large a population as possible. Only by using a large scale can an impoverishment

Table 1. Effect of mass selection on one gene. Portion of undesired types (percent) in selection on recessive (r) and dominant (d) characters (After Kappert 1953)

		F_2	F_3	F_4	F_5
Self-fertilization	r	75.0	0	0	0
	d	25.0	16.7	10.0	5.5
Cross-polination					
Selection *prior* to flowering	r	75.0	0	0	0
	d	25.0	11.1	6.3	4.0
Selection *after* flowering	r	75.0	50.0	25.0	12.5
	d	25.0	16.7	12.5	9.6

of different alleles of the gene loci be prevented, which could lead to inbreeding defects and reduced adaptability.

2.1.1.2 *Heritability and Response to Selection*

The response to selection can also be predicted within certain limits for quantitative characters that are due to a multitude of genes (polygenes). The success of selection with regard to the first generation after selection is generally termed R. It can be estimated according to the formula: $R = i \cdot h^2 \cdot \sigma$. Here, i stands for the intensity of selection, which can be calculated by an approximation formula. It is easier, however, to draw the value for i from corresponding tables (Becker 1975). σ stands for the measurable phenotypic variance of the plant character or the seed stock in which selection is to be carried out; h^2 denotes the heritability of that character. Although heritability is a statistical parameter to characterize the effect of polygenes, it can be explained simply by using one single gene as an example. If it is assumed that a small population consists of 4 AA, 8 Aa, and 4 aa genotypes with intermediate interacting alleles, the phenotypes exactly represent the genotypic constitution if no modifying effects of the environment on the character are considered. This character may, for example, be the height of the plants: aa may cause a plant height of 0.20 m, Aa = 0.40 m, and AA = 0.60 m. The distribution of this population will be as shown in Fig. 1a. The breeder now decides to select all plants that exhibit a height greater than 0.50 m. In this case, he will obtain all those phenotypes that are homozygous dominant (AA). The mean of the population to be selected is 0.40 m; the mean of the selected fraction is 0.60 m and the mean of its offspring (first generation) is also 0.60 m. The response to selection in one generation can be calculated as follows: 0.60 – 0.40 = 0.20 m.

When environmental factors cause phenotypical deviations, as shown in Fig. 1b, the breeder who selects all plants higher than 0.50 m will record less success. The average height of the selected fraction (2 Aa and 4 aa) is 0.57 m and the mean of its progeny is 0.53 m. Therefore, the response to selection is only 0.13 m (0.53–0.40 m). Although the intensity of selection in these two models differs little, the effects of the environment are obvious.

The variance of the phenotypes to be selected in model I can be calculated as 0.0213; the variance of model II is 0.0267. The source of the variance of model I is called the genetic variance (V_G). In model II, the total phenotypic variance (V_P) includes the genetic variance and an additional variance due to the environmental effects (V_E). The relationship between genetic variance and the total phenotypic variance is defined as heritability: $h^2 = V_G/V_P$, where $V_P = V_G + V_E$. In other words, heritability represents the strength of polygenes in relation to the environment[1].

[1]If there is more than one environment involved, the response of the genotypes may be different. In that case, the additional variance of the interaction between genotypes and environment ($V_{G/E}$) must be taken into account.

In model I, the heritability is 1.00 or 100%; in model II, this parameter is 0.80 or 80%. Consequently, the response to selection in model II is less than in model I.

Usually the interaction between two different alleles (A and a) is not exactly intermediate. In most cases, one allele is more or less dominant (A) and the other one is recessive (a). If, for example, model I is modified in such a way that Aa genotypes and AA genotypes represent the same phenotype (height of all these plants = 0.60 m, which means complete dominance), the 1:2:1 population averages 0.50 m and its variance is 0.0320. Again, all plants that exceed 0.50 m will be selected. The mean of the selected fraction is 0.60 m, but the mean of its first generation offspring is markedly lower, due to segregating recessives that originate from selected Aa heterozygotes (Fig. 1c).

Although no environmental effects are involved, there will be a loss of response compared to model I. The reason for this decrease is obviously the dominant effect of allele A.

Therefore, the genetic variance should be separated into two components: the so-called additive variance (V_A) and the dominance variance (V_D): $V_G = V_A + V_D$. The additive variance is useful for selection problems only. In order to predict R, the quotient V_A/V_P, which is called heritability in the narrow sense, should be incorporated into the formula for the response to selection.

When there is more than one gene under consideration, interactions between additive and dominant effects of the genes broaden the genetic variation. The components of the genetic variance that represent the interactions are called epistatic effects. [In classical genetics this term has another meaning: one gene interferes with the phenotypic expression of another nonallelic gene (or genes), so that the phenotype is determined by the former. In quantitative genetics the term epistasis is used to refer to all nonallelic gene interactions (Rieger et al. 1976).] For some quantitative characters, epistatic effects are very weak and they can be neglected for practical usage.

Heritability depends upon the genetic material investigated and the different environments involved. Many of our cultivated crops are grown as single individuals, e.g., fruit trees; others (cereals, beets, vegetables, etc.) are usually grown in fields that are more or less densely sown or planted. The heritability of quantitative traits such as yield also depends upon the size of the plots in field experiments. Such trials with replications are appropriate for estimating heritability. The table of variance of a simple, block-design experiment exhibits three sources of variance: treatment (= genotypes or families), replications (blocks), and error. Since such experiments represent one environment only, the obtained data are not very reliable. It is much more effective to repeat such experiments for several years at different locations. The table of variance then becomes much more comprehensive, and additional components of variances due to interactions between families and years, families and locations, etc., can be estimated by simple algebraic calculation (Table 2). The following formula enables the breeder to estimate the value of heritability:

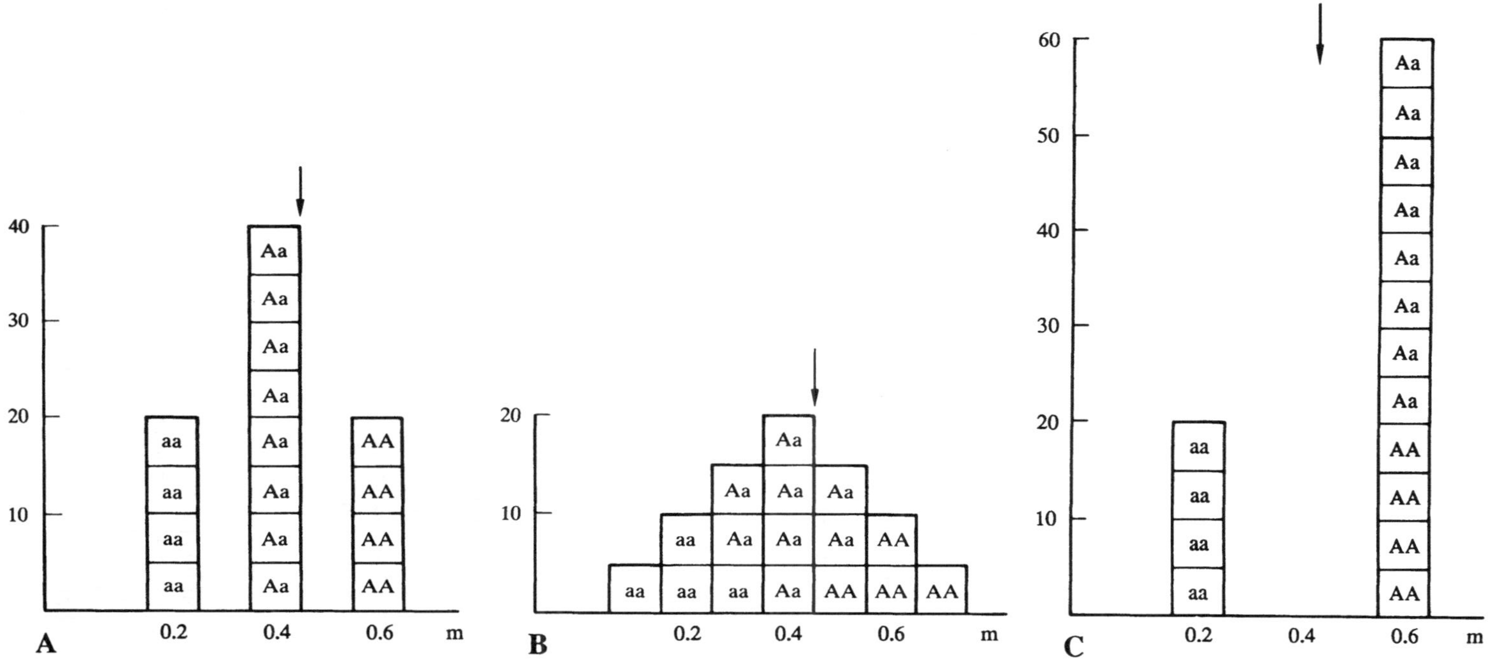

Fig. 1A-C. Frequency distribution of a segregating F_2 offspring. *A* Without any reaction to the environment; *B* with reaction to environmental influences; *C* with dominant effect of the A allele and without environmental effects; *Arrow* point of truncation (Schuster and Geidel 1978)

Table 2. Analysis of variance for data on genotypes (families) compared in replicated trials in several years and locations[a] (Allard 1966)

Source of variance	Degrees of freedom	Expectation of mean squares
Genotypes (families)	g-1	$\sigma^2_E + r\sigma^2_{GLY} + ry\sigma^2_{GL} + rl\sigma^2_{GY} + rly\sigma^2_G$
Genotypes × years	(g-1)(y-1)	$\sigma^2_E + r\sigma^2_{GLY} + ry\sigma^2_{GY}$
Genotypes × locations	(g-1)(l-1)	$\sigma^2_E + r\sigma^2_{GLY} + ry\sigma^2_{GL}$
Genotypes × years × locations	(g-1)(y-1)(l-1)	$\sigma^2_E + \sigma^2_{GLY}$
Error	(r-1)(gyl-1)	σ^2_E

[a] g,y,l, and r refer to the numbers of genotypes (families), years, locations, and replications, respectively.

$$h^2 = \frac{\sigma^2_G}{\sigma^2_G + \sigma^2_{GL}/L + \sigma^2_{GY}/Y + \sigma^2_{GLY}/LY + \sigma^2_E/RLY}$$

σ^2_G = variance of the genotypes (families)
σ^2_{GL} = variance due to the interaction of genotypes and locations
σ^2_{GY} = the genotype by year component of variance
σ^2_{GLY} = the genotype by location by year component of variance
σ^2_E = the error variance

G = number of genotypes
L = number of locations
Y = number of years
R = number of replications

There are additional methods for estimating the components of variance. If these values are available for a distinct region, it is possible to optimize the allocation of field experiments, i.e., the breeder can estimate how many replications, locations, and years would provide the best chance of detecting that family which promises to become a good cultivar. The heritability can also be determined by the correlation between parents and their offspring. But it is not the object of this book to describe all of these procedures in detail. Special publications are available in abundance (refs. in Wricke and Weber 1986).

2.1.1.3 Natural Mass Selection

Positive mass selection has proven to be very effective combined with *natural* selection in forage crops, grasses, and leguminous fodder crops. The good results that have been obtained in the use of natural genetic selection when breeding new forage crop varieties reveal an important principle of breeding, of general validity: breeding should, if possible, take place under those environmental conditions (climate and soil) in which later cultivation of the new variety is intended. Further, the selected material should be tested using the same cultivation technique as intended for the later variety in practice. In the perennial cultivation of forage crops, especially in preparing meadows and pastures, a mixture of several species is often used. Due to narrow, regular planting, competition occurs not only between genotypically different plants of the same species, but also between plants of different species. Therefore, individual grass plants are not only cultivated separately in the nursery, but also often within a crop of another grass species, in order to observe their behavior under competitive conditions likely to occur in practice.

When an introduced population of a cross-pollinated perennial forage plant is exposed to natural selection at its new location, the success of that natural selection is strongly dependent on the timing of harvest, from which the following generation shall be grown for further selection. If seed is already harvested in the 1st year of cultivation, a population or variety with early developing shoots will form, showing its highest yield in green matter within the first growing period and markedly falling off in performance in the following years. If seed is not harvested until the 3rd year of cultivation, a variety will develop consisting of strongly persevering types, decreasing in yield of green matter in the 1st year, but with a higher and more stable yield in the following years. Its overall performance is consequently more valuable if forage crop cultivation is planned for several years. This demonstrates that natural selection and selection by the breeder can be effectively coordinated.

In *cross-pollinated*, cultivated species mass selection, including natural selection, will succeed in developing types or mixtures of types that are characterized by particularly good adaptability to the corresponding environmental conditions and, therefore, will also result in reliable and constant yields. In other features, however, including those of commercial use, they may still display a more or less strong variation. For formal reasons, e.g., a registration of varieties and the variety protection connected therewith, this remaining variation should be no more restricted than absolutely necessary. The resulting breeding work is not only uneconomical, but often even detrimental, since too strong a limitation of this remaining variation and, consequently, the loss of gene alleles may sometimes lead to decreased performance.

In crops of *self-fertilizing*, cultivated plants such as wheat, barley, oats, peas, beans, and others, natural selection can also change their genotypic composition. In countries with extensive agriculture, such crops can often still be found in cultivation. They are the result of long-standing selection processes and are, therefore, well adapted to the conditions of their environment. They are characterized by more or less constant, but not maximum, yield performance. If such populations are transferred to other locations or if their cultivation methods are changed, for example, by application of commercial fertilizers, irrigation, or plant protection measures, the genotypic composition of the population in question will change as well. Since self-fertilizing populations mainly consist of homozygous individuals, the result will be a numerical shift of the individual genotypes or even the total elimination of some genotypes, but not so much in recombination of genes leading to entirely new genotypes, as is the case in outbreeders. Therefore, the genetic potential of adaptability of cross-pollinated crops is greater than that of self-fertilizing crops.

How strongly natural selection can also affect self-fertilizing crops is demonstrated in experiments with artificially produced variety mixtures of barley. In an experiment in the USA, 11 barley varieties were mixed in equal portions and cultivated at ten different locations under very different environmental conditions for 4-12 years. As can be seen in Table 3, the variety "White Smyrna" completely disappeared from the mixture when cultivated in Ithaca (NY), whereas in Moro (Oregon) "White Smyrna" almost entirely superseded all other genotypes of the mixture (Harlan and Martini 1938).

Natural selection is not only an effective tool for the breeder, it is also an important factor that must be considered in multiplication, i.e., in seed production for the market, in order to prevent undesired changes in the genotypic composition of a variety.

Table 3. Effect of natural selection on a mixture of 11 barley varieties after 4–12 years of cultivation in different areas (After from Harlan and Martini 1938)

Cultivar	Areas Arlington, Va.	Ithaca, N.Y.	St. Paul, Minn.	Fargo, N. Dak.	N. Platte, Neb.	Moro, Ore.	Davis, Calif.
Coast and Trebi	446	057	083	156	224	006	362
Gatami	13	9	15	20	7	0	1
Smooth Awn	6	52	14	23	12	1	0
Lion	11	3	27	14	13	0	8
Meloy	4	0	0	7	4	0	27
White Smyrna	4	0	4	17	194	489	65
Hannchen	4	34	305	152	13	4	34
Svanhals	11	2	50	80	26	0	2
Deficiens	0	0	0	1	3	0	1
Manchuria	1	343	2	37	1	0	0

Sum of evaluated plants = 500 each

2.1.2 Pedigree Breeding of Self-Fertilizing Crops

The nature of the pedigree method lies in the cultivation of a *separate* offspring of each plant selected from a population. The value of the selected plant is ascertained by progeny tests. Thus, a selection of genotype takes place, whereas mass selection is principally a selection of phenotypes. This difference between the two breeding methods becomes especially obvious if a character with monogenic-dominant inheritance, e.g., stringless pods in dwarf beans (FF or Ff = stringless, ff = stringy), is selected from a segregating generation of a self-fertilizing crop. If stringy plants (ff) are rogued in the F_2 and if they are discarded by mass selection in the following generations, stringy plants will still segregate in the F_3, F_4, and F_5 at portions of 16.7, 10, or 5.5% (Table 1), respectively. Although the number of undesired types decreases, no complete breeding success is accomplished. If, on the other hand, separate F_3 progenies are cultivated from a greater number of stringless F_2 plants (FF or Ff), very probably one-third of all progenies will be constantly stringless. The breeding aim concerning this character can therefore already be realized in the F_3. When selecting dominant characters in a population of self-fertilizing crops, selection of individual plants with progeny tests is definitely superior to mass selection.

Although individual selection was already applied purely intuitively in the second half of the last century by ingenious plant breeders who had no knowledge of Mendel's laws, it was not widely practiced and successfully applied until after the rediscovery of Mendel's laws. In wheat, barley, oats, peas, and other self-fertilizing species, populations can be divided into lines and these can be tested for performance so that new, valuable varieties can be created.

When a local variety of a cereal species is to be improved by breeding, a greater number of plants (100–1000) whose phenotype is closest to the set breeding aim are selected from the field crops. Previously, these selected, elite plants were extensively weighed and measured for stalk length, ear length, ear density, tillering, and 1000-kernel weight. These evaluations were based on the belief that certain exterior characters are correlated with yield ability. However, since correlations are rarely complete and the formation of quantitative characters quite easily varies depending on environmental conditions, the merit of weighing and measuring elite plants is restricted. Extensive series of examinations for *qualitative* characters on individual plants are only justified if the presence or absence of a substance is to be tested, e.g., alkaloid content or absence of alkaloids in lupines. The progeny of a selected, elite plant is usually termed A line or A family. Sowing takes place in rows or in small plots. Evaluation of the A lines is done mainly by eye inspection, since quantitative ascertainment of yield does not produce reliable results because of the small size of the plots and the lack of replications. In order to have a better standard of comparison, so-called standard plots containing the local variety or other modern cultivars are interspersed with the A lines at specific intervals.

After close observation and evaluation during the vegetation period, 80–90% of the A lines are excluded from further breeding. The remaining lines are harvested separately and are entered in the next generation, the so-called B line trial (Fig. 2), in the following year. In this trial, the individual lines can be sown in larger plots arranged in replicates, so that quantitative yield determination can be done for the first time. Based on the trial results, 10–20% of the B lines is determined for the C line trial in the following year, which is in principle accomplished in the same way as the B line trial, except that the plots are larger and the number of replications and comparison controls are increased. After the C line trial, sufficiently reliable statements can be made as to whether or not the performance of a line is superior to the control. For confirmation of the results, a D line trial will ensue, which should also simultaneously be arranged at other locations, in order to obtain an insight into the ecological variance of the new

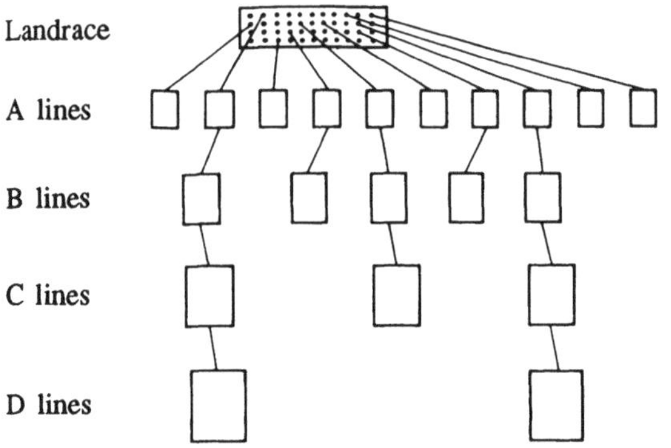

Fig. 2. Procedure of selection in a self-fertilizing landrace

breeding lines. At the same time, the remaining seed of the best lines is multiplied in the field so that, after passing the official test and obtaining the license for the new breeding line as a protected variety, there will be sufficient seed available for the market.

2.1.3 Pedigree Breeding of Cross-Pollinated Crops

Since in *self-fertilizing crops* the ovules of a plant are fertilized by the pollen of the *same* plant, the hereditary constitution of its offspring is determined in equal parts by the hereditary constitution of the ovules and by the pollen of *one* female parent. Thus, in self-fertilizing crops selection takes place according to the hereditary value of the female as well as the male parent. Both parameters are known and are recorded in the same way in progeny tests after pedigree breeding. In *cross-pollinated crops*, on the other hand, selection takes place chiefly according to the hereditary value of the female parent. The male part is always unknown, consisting of the pollen mixture of a multitude of plants, and all plants of a field crop have the same prospect of being pollinated by this mixture. If differences occur between the progenies of individual plants of an outbreeding species, it is assumed that these are mainly related to differences within the *female hereditary portion*, since quite probably they have all been pollinated by the same pollen mixture. Therefore, the task of breeding in cross-pollinated crops is not only to discern the hereditary value of the female parent by selection, but also to improve the unknown factor of the male parent by suitable methods, i.e., to succeed in *regulating cross-pollination*. The possibilities of regulating cross-pollination are mainly dependent upon whether economical use of the species takes place *before* or *after* flowering; furthermore, they depend on the reproductive biology, whether or not there is any seed set after forced self-pollination, and how far inbreeding depression occurs within the offspring of selfed plants. The potential of vegetative propagation is also of importance for the choice of selection methods.

Regulation of cross-pollination can be managed quite simply in species that are used commercially *before* they flower. This includes, e.g., spinach and radish in the annual species, sugar beets, carrots, and cabbage in the biennial species, and many perennial grass, clover, and lucerne species. The fact that it is possible to ascertain the performance value of these species in the A, B, and C trials before flowering means that all deleterious families can already be rogued before flowering. Consequently, the genetic composition of the pollen mixture is improved; mutual fertilization only takes place between families that are above average. The breeder should also take advantage of the possibility to discard, before flowering, individuals deviating from the standard in families that are destined for further breeding. It is not always necessary to exclude above average families from further breeding because of single, deviating individuals.

Regulating cross-pollination is more difficult in species whose economic value can only be determined by examining, *after* flowering, the product in which the grower is interested, e.g., rye, hemp, or cucumber. In rye breeding, e.g., the method of overstored seed has proved effective for regulating cross-fertilization. First, a greater number of individual plants is harvested from a mixture of types of a population, and their progenies are sown as A families in smaller plots in the same way as in selection breeding of self-fertilizing crops, except that half of the seed of all elite plants is retained in reserve. Those A families meeting all requirements of the breeder are not directly multiplied as in self-fertilizing cereals, but the remaining seed of the corresponding elite plants is sown in the following year at as great a distance as possible from the other crops; it then enters a so-called A' family trial. The reason for this is that since the economically valuable characters of the rye A families can only be definitely evaluated after maturity, the best A families destined for further breeding have already been pollinated by a pollen mixture that also contains pollen of less valuable lines. Therefore, the hereditary nature of the A families for further breeding has most probably deteriorated again. Thus, if sowing of only the selected A families is repeated using the overstored remainder of the seed, these lines will be fertilized by an above average pollen mixture because the pollen of the inferior A lines has been eliminated (Fig. 3). Consequently, simple regulation of cross-pollination, which considerably improves the pollen mixture, is achieved by repeated cultivation of the best A families. The A families are only a preselection; the seed for the B family trials is first obtained from the A′ families. The setup of the A′ family trials and their evaluation is carried out in the same manner as that of the A families. The seed for the B family trial is also split and one-half is overstored in order to subject the best B families to a B′ family trial. Using the overstored seed, the best B families are subjected to a B′ family trial in the following year. They are carried out in the same manner as the A′ family trials: all inferior progenies are eliminated before flowering insofar as they can be identified as such. If the B′ family trial has not provided clear information on the value of the individual families, a C or C′ family trial must follow. Otherwise, the seed of the best B family is multiplied in the field. Equivalent families can also be multiplied together.

When evaluating and selecting families for further pursuance, the breeder must realize that evaluation and selection are only based on the weather conditions predominating in the corresponding year. Thus, a different evaluation of

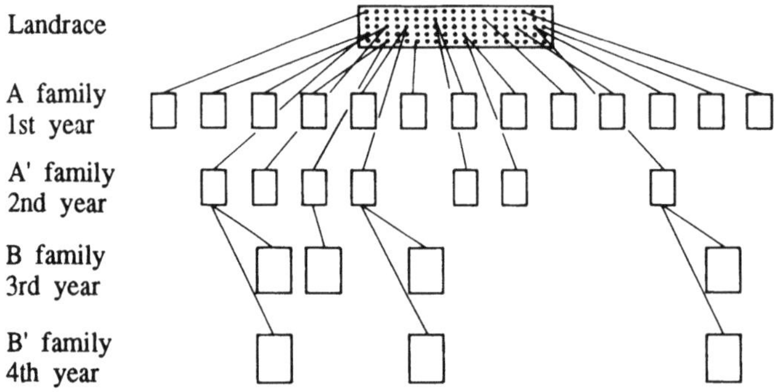

Fig. 3. Procedure of selection using overstored seeds

the test families for drought resistance is only possible in extremely dry years, and differences in resistance to rust can only be determined during a severe epidemic. This is the reason for the failure of lines that have proven successful in the A and B trials, when evaluated in the A′ or B′ family trials of the following year. They must, therefore, be excluded from further breeding. In order to be able to evaluate the reaction to as many different environmental conditions as possible. The breeding material must be examined at different locations.

Culling entire families before flowering also leads to a selection on the hereditary value of the male parent; the pollen involved, however, always remains a mixture of different genetic constitutions, since it is produced by several lines. Only if outbreeding crops are artificially forced to self-fertilize is the origin of the male hereditary material within the offspring of a plant exactly known and, thus, the control of pollination is achieved.

Breeding of cross-pollinated crops can be considerably facilitated and intensified if vegetative propagation of the species in question is possible, e.g., by repeated division of a plant in perennial grasses, clover species, and lucerne. Employment of vegetative propagation in the breeding cycles offers the following advantages: if clones are produced by vegetative propagation before the progeny trials of individual plants (A family trial), these clones can be more reliably judged in regard to their valuable characters, as is the case in individual plants. Furthermore, regulation of cross-pollination can be carried out in clone trials by eliminating the unfit clones before the trial, so that better progenies can be expected from clones than from individual plants. The increased seed production of the clones permits the breeder to carry out the following trials on a wider scale, i.e., with larger plots and more replications. With this, the breeding cycle is accelerated and the reliability of the results increases.

2.1.4 Selection in Vegetatively Propagated Crops

For commercial use, some important cultivated plants such as potato, sugar cane, many fruit species, and ornamental plants are not multiplied by seed, but are vegetatively propagated by tubers, cuttings, grafting of scions on root stocks, etc. All plants of a vegetatively propagated variety are identical in a genetic respect, i.e., they form a *clone*, which is one individual divided into many parts with the same genetic constitution. Therefore, no genetic variation is to be expected in a clone variety, unless mutations, i.e., changes of genotype, occur spontaneously within the somatic tissue, resulting in changes of morphological traits or physiological behavior. Such sports, as these somatic mutations are called by growers, have been important for the development of new varieties, especially in fruit species and ornamental plants, and they also became of interest in potato breeding, especially in the USA. But first to the selection method itself. How does a potato breeder, for example, proceed in order to create a new variety with specific combinations of characters? Populations consisting of a mixture of different clones are not available in our degrees of latitude. They are only distributed in Mexico and South America, the habitat of the cultivated potato. However, they cannot be taken into consideration as original material for the selection of new varieties, because they are primitive forms of the cultivated potato and of related wild species, lacking in important traits.

New extensive genetic variation for selection can only be generated by raising potatoes from seed. Since potatoes are strongly heterozygous, the first progeny of one female plant or from a cross of two genotypes is a segregating generation consisting already of a great number of different genotypes. It can be compared to the F_2 of a self-fertilizing species, in which a heterozygous but uniform F_1 generation results from a cross between two homozygous parents. Each genotype that is selected from a segregating generation is constant in vegetative multiplication. Breeding for uniformity and stability, which is especially difficult in cross-pollinated species, can be omitted. With successful selection of a valuable genotype, breeding success is immediately established. This applies for all vegetatively propagated cultivated plants.

The first major difficulty in breeding vegetatively propagated species lies in the production of seedling populations that contain as many desired variants as possible, and furthermore in the development of selection procedures that help to quickly and reliably find these variants. Therefore, breeding must begin with the selection of suitable parents for breeding seedlings and subsequent seedling selection. Since potatoes are strongly heterozygous and the result of recombination of the genes during gamete formation in the parent plants regarding the formation of new geno- and phenotypes cannot be predicted, a selection according to phenotype of the parent plants is not sufficient. Practical experience has shown that some good varieties form preponderantly inferior seedling progenies, just as very valuable genotypes can sometimes be found in the progenies of poorer varieties. Consequently, the genotypic constitution of the parents or their genotypically conditioned combining ability with regard to the production of offspring having a desired genetic constitution must be ascertained in a different way.

For this purpose, two methods have been developed. In one of the methods, ca. 1000 seedlings derived from selfings of every variety to be examined are cultivated. The proportion of desired phenotypes among them is ascertained, e.g., the proportion of types with yellow flesh or round tubers or of types with extremely early maturity. Since the progenies of selfed plants usually display strong inbreeding depression, generally no varieties can be developed from these inbred (I) progenies. If two of these varieties analyzed by inbreeding are crossed, the proportion of desired types to be expected can be estimated by the mean of the investigated characters in the selfed progenies of both parents.

With the second method, the general combining ability of varieties is determined by crossing them with a known test parent. The cultivation of about 300 seedlings from each cross suffices for this analysis. The advantage of this method is that selection can already begin in the progeny of a testcross, if this proves to be useful; thus, time and work are saved in the breeding process. In the normal process, however, the parents of the best progenies are first crossed with one another and then selection of individual plants takes place in their progeny.

Due to pricking-out, potting, and transfer to the field, the seedlings are often subjected to very different treatments and environmental conditions, so that the expressions of the characters are often strongly modified towards either the positive or negative side. Since selection of seedlings grown under these different conditions is not promising, a so-called bulk generation is introduced before selection. This is done by taking one tuber from each seedling of a progeny and planting these together as a bulk population in the following year. From this population, plants suitable to the breeding aim are selected to be tested as A clones (8–10 plants each) in the following year (Fig. 4). According to field observations, about 80–90% of the A clones are eliminated. The remaining clones are entered in a B clone trial in which more exact examinations can be made. Statements on the breeding success are only possible after C and D clone trials, the trial elements of which constantly decrease by further elimination of unsuitable genotypes. The trials are also carried out with replicates and at different locations.

In practical breeding, the methods must be modified according to the individual plant species, especially as regards their different reproductive biology. Thus, for example, a parent selection of kernelled and stone fruit, by testcrosses and progeny analysis, is necessary. The selected seedlings, however, must be propagated by grafting scions or budding on stocks for further trials. Breeding new varieties in kernelled and stone fruit with the help of seedlings is made very difficult by the long generation intervals: determination of the hereditary value of the cross parents and of the economical value of a seedling selected from the cross takes 6–7 years each. Therefore, new selection procedures are being developed for species with long generation intervals which will be described in Chapter 3.

All vegetatively propagated species have the great disadvantage of easily spreading pests and diseases, especially virus diseases, from one generation to the next by vegetative propagation. In cases of strong infection, this may then cause complete breakdown of varieties. Since the majority of known virus diseases is not transferred by seeds, the question arises as to how far breeding is capable of creating varieties of vegetatively propagated species growing true to type. References to this can be found in Sections 3.3 and 3.4.

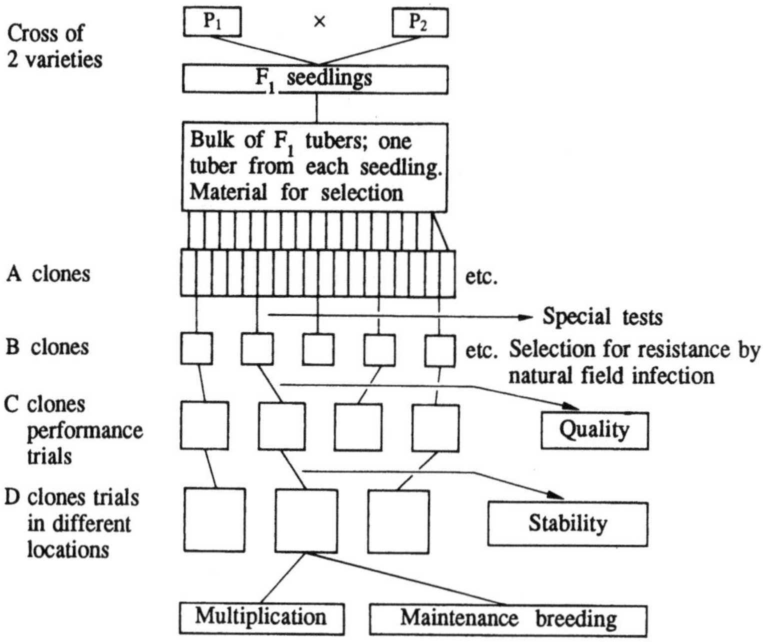

Fig. 4. Breeding new varieties in potatoes

2.2 Combination Breeding

If the variation in the available landraces, primitive forms, or ecotypes has been exhausted, new genetic variation for further selection can be obtained by crossing different genotypes of a species. As a consequence of a cross between different genotypes, a recombination of the alleles of the genes takes place in the F_2 and subsequent generations. Due to this recombination of the alleles, the following important events can occur:

1. The recombination of the alleles has, as a consequence, the recombination of the characters controlled by the genes. Characters that are economically valuable and are distributed among both parents can be combined within a new genotype and later bred to constancy.

2. By crossing, genotypes with new characters can occur in the F_1 and F_2, which had been unknown in both parents. They result from the interaction of genes deriving from both parents. These new characters can also be made homozygous.

3. Also, plants with a change in quantitative characters may occur after the cross. They may transgress the degree of the parental features in the intensity of expression of character either towards the positive or the negative side. The crossing of parents marked by extremely positive or negative expression of a quantitative character may lead to genotypes with a constant intermediate expression of the character concerned.

2.2.1 Genetic Bases of Combination Breeding

The genetic bases of combination breeding were established more than 100 years ago by Mendel, who was the first to realize, by exact determination of the different plant types within the F_2 progenies, that single, heritable factors are responsible for the expression of specific traits. These factors, termed "genes" by Johannsen in 1909, proved to be more or less easily combinable with each other. However, it was soon realized that many features cannot simply be attributed to the effect of only one gene or a few heritable factors. Usually, these include the very characters of the cultivated plants which, from the human point of view, should most particularly be improved by breeding, such as yield capacity, resistance to extreme climatic conditions, content of specific plant products, etc. It could be proven, however, that the expression of such quantitative features is also controlled by genes, the number of which cannot generally be precisely determined. However, just like the genes responsible for the qualitative characters, these behave according to the laws of Mendel.

As a rule, genes with qualitative effects can be clearly estimated by suitable cross experiments. For example, if two homozygous plants, whose genetic constitutions differ in two factors only, are combined the F_1 generation will exclusively consist of double heterozygous genotypes, because the cross AA bb × aa BB as well as the cross AA BB × aa bb will always result in Aa Bb. The genotypic segregation in the F_2 generation can easily be determined using the chessboard pattern by Punnet, once the gametic segregation has been entered in the appropriate table headings (Table 4). In the gynoecium as well as in the androecium of the F_1 hybrids, gametes of the constitution AB, Ab, aB, and ab are formed in equal quantities. The genotypic segregation to be expected of such a bifactorial cross combination of the F_2 descendents will therefore appear as follows:

1 AA BB
2 AA Bb
1 AA bb

2 Aa BB
4 Aa Bb
2 Aa bb

1 aa BB
2 aa Bb
1 aa bb

16

Phenotypically, this segregation ratio can only be found if both genes have an intermediate effect. Only then can the heterozygous types (Aa and Bb) clearly be distinguished from the homozygous ones. In many cases, however, one allele of a gene is dominant over the other allele, the latter being termed recessive, so that the phenotypic segregation will simply be 9 A • B • : 3 A • bb : 3 aa B • : 1 aa bb. The way of writing either A • or B • indicates that the class of phenotypes

Table 4. Chessboard pattern for determining the genotypic segregation of the F_2 generation concerning two genes

♀ \ ♂	AB	Ab	aB	ab
AB	AA BB	AA Bb	Aa BB	Aa Bb
Ab	AA Bb	AA bb	Aa Bb	Aa bb
aB	Aa BB	Aa Bb	aa BB	aa Bb
ab	Aa Bb	Aa bb	aa Bb	aa bb

affected includes the homozygous dominant AA or BB, as well as the corresponding heterozygous Aa or Bb. Depending on the constitution of the cross parents, certain genotypes prove to be new combinations in the F_2.

If, for instance, AA bb and aa BB are crossed, genotypes AA BB and aa bb appearing in 1/16 of all cases are formed. Due to the double homozygosity, no segregation is to be expected in their progenies (F_3).

In cases where the crossed parents differ in three characteristics, each controlled by one gene, the much larger multiplicity of types to be expected can be recognized by the same conformity. If, for instance, in lupines one parent has the genetic constitution AA bb CC [A = hard and a = soft skin, B contains bitter substances and b is free of bitter substances (sweet), C = dark seeds and c = light seeds] and the other parent has the constitution aa BB cc, the hybrid Aa Bb Cc will segregate into eight classes of phenotypes in the F_2 generation. The frequencies of the phenotypic classes of such a trifactorial inheritance may be derived from the three monohybrid segregations and can be seen in Table 5.

Besides the two parent classes (P), there are six new combinations (NC) that will not segregate any further if they are homozygous for all gene loci.

The phenotypic segregation ratio of the F_2 is not only determined by the different effects of the alleles of a gene (intermediate-dominant). The genes can also mutually influence one another. These gene interactions result in F_2 segregations distinctly differing from the norm (9:3:3:1 if two genes are involved).

Certain substances can only be produced by a plant if two (or even more) genes come together in a specific allelic constitution. For example, the violet color of the blossoms in gillyflower (*Matthiola incana*) is only produced if the nuclei of the plant in question contain both the dominant allele A of one gene *and* the dominant allele B of the other gene. The genotypes AA bb and aa BB cannot produce the color and they therefore have white flowers. Only the cross product (Aa Bb) will show the *complementary* effects of the genes; the F_1 hybrid has violet flowers. One-sixteenth of the plants in the F_2 progeny are homozygous AA BB and are therefore a new form of gillyflower, the progenies of which will not segregate any further.

Table 5. Segregation of three genes (for details, see text)

3 hard	3 bitter	3 dark =	27 hard	bitter	dark	A·	B·	C·	(NC)
		1 light =	9 hard	bitter	light	A·	B·	cc	(NC)
	1 sweet	3 dark =	9 hard	sweet	dark	A·	bb	C·	(P)
		1 light =	3 hard	sweet	light	A·	bb	cc	(NC)
1 soft	3 bitter	3 dark =	9 soft	bitter	dark	aa	B·	C·	(NC)
		1 light =	3 soft	bitter	light	aa	B·	cc	(P)
	1 sweet	3 dark =	3 soft	sweet	dark	aa	bb	C·	(NC)
		1 light =	1 soft	sweet	light	aa	bb	cc	(NC)

New combinations can also be formed in the F_2 if, in a cross parent, the visualization of the effect of one of the genes is prevented, due to the simultaneous presence of two independent genes. Only after they have been separated by a new combination of the genes in the hybrid during formation of the gametes, are genotypes formed possessing only one independent gene and therefore showing the new character. Segregations of this kind occur in the F^2 of two dwarf bean varieties with black or white beans, respectively. The hybrid has black beans and in the F_2 there is a segregation of 12 black : 3 brown : 1 white (Fig. 5). From this it can be concluded that the black color of the bean is due to a gene S and the brown color to a gene B. The S·B· genotypes are also black, because the effect of S covers up the effect of B, as can be seen in the pattern of Fig. 5.

This is a case of *dominant epistasis*: S is epistatic over B or B is hypostatic to S. There are also new black genotypes SSbb in the F_2 that do not segregate any brown types after a cross with white ssbb beans, but result in a simple monohybrid segregation of 3 black : 1 white. The effect of recessive alleles of a gene suppressing the phenotypic manifestation of another gene is called *recessive epistasis*. The interaction of genes that are combined by crossing parents of different genetic compositions can result in stronger or weaker expression of characters or features. Thus, there is a *quantitative* change in *one* characteristic only. This is different from the complementary effect of genes described above, which is qualitatively expressed by formation of *new* characteristics or features. The F_2 of crosses of this kind have a continuous variation, which a Mendelian analysis, by forming classes of characteristics that can clearly be distinguished from one another, does not make possible. It is not seldom that the variation of the F_2 in the examined material exceeds the extent of the variation of both cross parents, which is termed transgression. Continuous variation is controlled by Mendelian genes just as is discontinuous variation, which is evident after analyzing the way the grain color of wheat (*Triticum aestivum*) is inherited. After crossing two wheat plants having medium red grains, the F_2 may segregate 15 red : 1 white, i.e., the

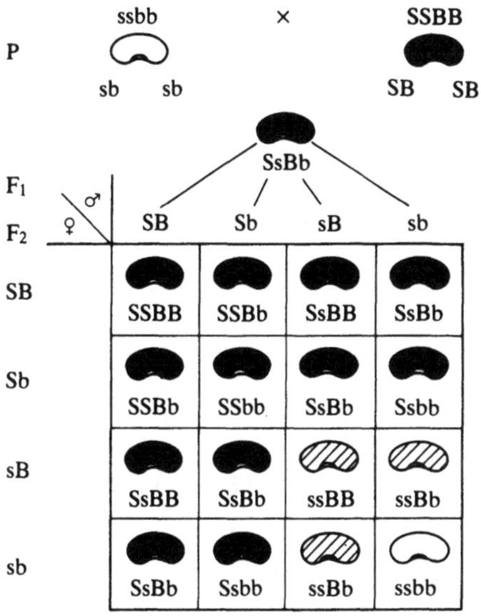

Fig. 5. Cross between two dwarf bean varieties with black or white beans demonstrating epistasis

cross parents differ in two genes that are involved in the segregation of red : white. The plants with white grains appearing at a frequency of 1/16 in the F_2 generation are the double recessive genotypes. The 15/16 plants with red grains vary strongly in the intensity of the red color, covering all degrees from light red, medium red, and red to dark red, i.e., the parents' light as well as dark red grain colors are transgressed. Since both genes have the same phenotypic effect, they are termed R_1 and R_2, so that both medium red parents have the genetic constitution $R_1R_1r_2r_2$ or $r_1r_1R_2R_2$, respectively, and their hybrid, $R_1r_1R_2r_2$. Assuming that the red color becomes stronger with an increasing number of dominant alleles, but becomes weaker when they are missing, then all genotypes with four dominant alleles are dark red, with three dominant alleles red, two dominant alleles medium red, one dominant allele light red, and with no dominant alleles white.

If the genotypes resulting from a bifactorial segregation are entered in a Punnet pattern, the square is rotated as shown in Fig. 6, and if the frequencies of individuals of the five phenotypic classes white, light red, medium red, red, and dark red are projected onto the base line of the figure, a frequency ratio of the five phenotypic classes of 1:4:6:4:1 can be found. Thus, their distribution is characteristic for quantitative features with continuous variation. From this it can be recognized that a continuous variaton, which transgresses in its extent that of the parents, can also be due to genes in which the parents differ. Genes affecting the formation of a quantitative feature in the same way are termed polygenes or polymeric genes. The effect of a single polygene on the phenotype is generally small. On the other hand, environmental factors have a strong modifying influence on the formation of features, so that different genotypes can resemble each other in phenotype. A survey of some gene interactions and their effect on the phenotypic segregation of the F_2 is given in Fig. 7.

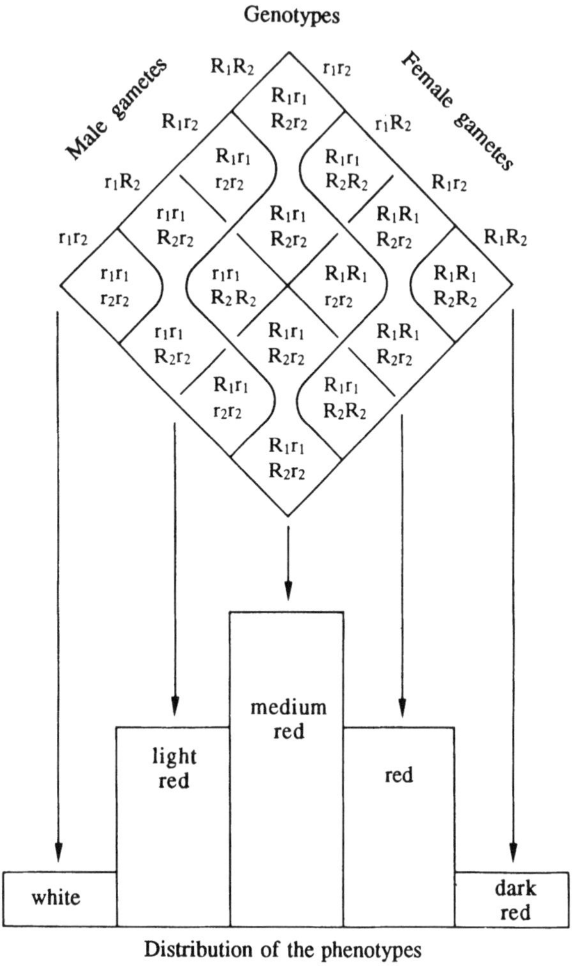

Fig. 6. Segregation of grain color in wheat

It becomes evident that the breeder's definition of a gene is a mathematical phrase giving a measurement for phenotypic segregations based on the frequency of crossovers documented on a hypothetical map: the Morgan gene map. For the elaboration of this map, no information on the biochemical nature of a gene is necessary. In addition to this definition of a gene, the molecular analysis of the chemical composition of the genome's physical basis is developed. From molecular genetic research we know today that the genome is – with some exceptions, e.g., most plant viruses – desoxyribonucleic acid (DNA). The most important feature of DNA is that is usually consists of two chains twisted about one another in the form of a double helix (Fig. 8). Each chain is a polynucleotide

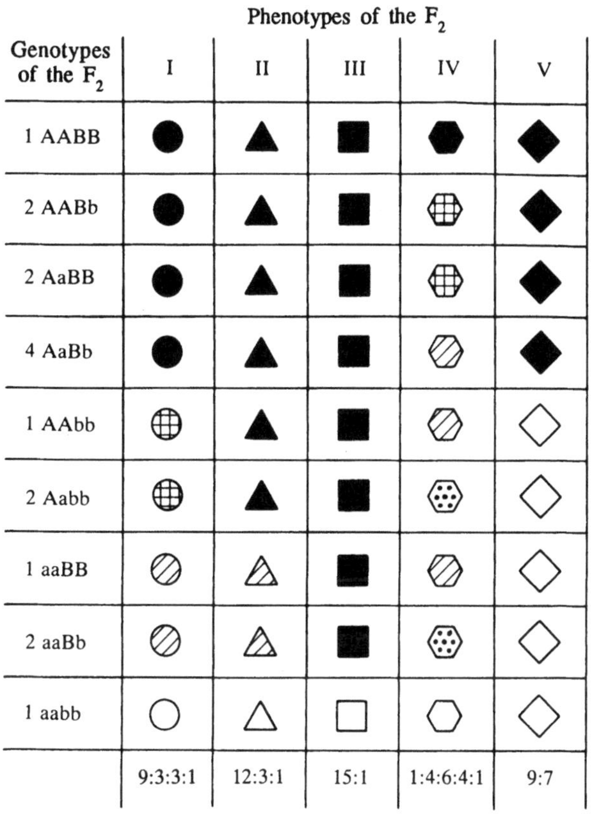

I A is dominant over a, B is dominant over b (no interaction)
II A is epistatic over B and bb (epistasis)
III A is equidirectionally replaced by B and vice versa
IV The dominant alleles A and B behave additively in the same direction, AA Bb = AaBB, Aa Bb = AA bb, etc.
V A character can only be phenotypically expressed, if A and B are present (complementary effect)

Fig. 7. Genotypic and phenotypic segregation of F_2 populations involving two interacting genes (Schuster and Geidel 1978)

in which the sugar of each nucleotide is linked by a phosphate group to the sugar of the adjacent nucleotide. Four main nucleotides exist, each of them containing a deoxyribose, a phosphate group, and a purine or pyrimidine base. There are two purines: adenine (abbreviated A) and guanine (G) and two pyrimidines: cytosine (C) and thymine (T). The two chains are joined together by hydrogen bonds between base pairs in such a way that adenine is always paired with thymine (A-T or T -A) and cytosine with guanine (C-G or G-C). The sequence

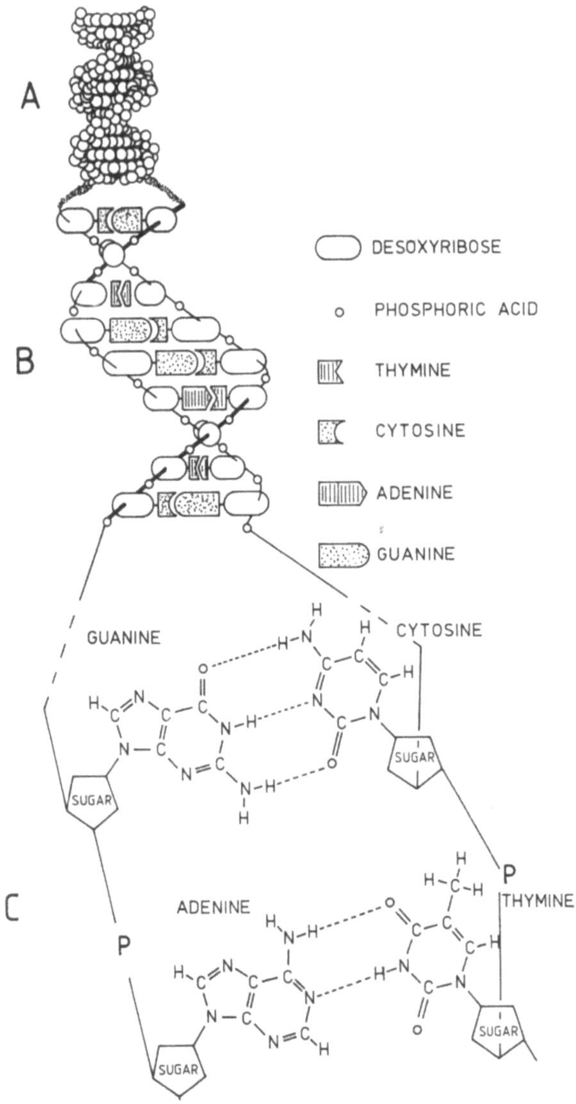

Fig. 8A-C. Watson-Crick model of desoxyribonucleic acid (DNA). *A* Double strand; *B* chemical construction; *C* specific base pairing

of the base pairs is the key for the genetic information. Three base pairs code for one amino acid; several or many amino acids code for one enzyme, and many enzymes finally make the phenotype.

As will be described later, we are witnessing today the translation of the breeder's gene expression into the molecular scientist's expression and vice versa. First phenotypes can be described as a nucleotide sequence such as, e.g., the seed storage proteins of barley, the hordeins. Classical and molecular genetics are becoming a unit. For almost all applied breeding programs, the classical approach is the one used today and thus we will concentrate predominantly on the description of classical breeding methods.

2.2.2 Combination Breeding Procedures

The first prerequisite a breeder must have to succeed in combination breeding is a good knowledge of the existing varieties and, furthermore, an insight into the genetic variability of the species in question. Only with this experience will he be able to select, with any degree of certainty, the suitable cross parents. Since genetic analyses of the traits important for breeding are only available for a few species, conclusions about genetic breeding behavior cannot simply be drawn from the phenotype of a plant. Moreover, nonpredictable new formations, resulting from the interaction of the genes distributed among the two parents, can occur in the subsequent generations; transgressions in particular can seldom be safely predicted. Thus, the breeder must equally expect unpleasant as well as pleasant surprises. It is therefore recommended that an optimum number of crosses be carried out. If an existing variety is to be improved in one or more valuable traits, it must be crossed with several varieties having the desired characters. As the field area is limited, it is recommended first to test a larger number of combinations with fewer individuals in the segregating generations rather than to test fewer combinations with many individuals. If necessary, the cultivation of promising F_2 segregations may be repeated with a larger number of individuals.

Important consequences for the practical implementation of combination breeding arise from the propagation system of a species, i.e., whether it is self- or cross-fertilized. The genotypes of a self-fertilizing crop used for a cross are largely homozygous, as far as this can be inferred from their phenotypical behavior. This means that their cross will result in a homogeneous heterozygous F_1 generation. Since it is not always certain that the F_1 generation exclusively consists of true hybrids and does not also contain single plants from selfing – in which the cross failed because of too late an emasculation – it is recommended that the F_1 plants be separated for further cultivation, so that F_2 progenies that do not segregate can be excluded from further breeding. If one of the parents used in the cross has a character with dominant inheritance, this genotype should be used as male cross parent since, if the cross is successful, all F_1 plants must display this dominant character. If the cross parents are not homozygous, homozygous lines can be obtained in self-fertilizing crops by self-pollinations within the progenies. The same also applies for optional cross-pollinated crops, such as yellow lupines, field beans, rapeseed, and others.

2.2.2.1 Combination Breeding of Self-Fertilizing Crops

If it is the breeder's intention to begin selection of new genotypes and their progeny tests already in the F_2 generation, it is recommended that the F_2 not be cultivated on too small of a scale, so as to be able to realize a large number of new combinations. The larger the number of genes in which the crossed parents differ from one another, the smaller the prospect is of obtaining the desired combinations already in the F_2, especially if they are caused by recessive genes. Most of the F_2 plants will be heterozygous dominant, so that the majority of their progenies (A lines) will segregate. For this reason, the number of individuals of the A lines should not be too small, so as to accomplish many combinations and to allow an insight into possible segregation ratios. Among the A lines, the breeder will seldom find any plants that already correspond to his breeding aim and that are at the same time constant. Therefore, he must again select new "elite" plants from the best A lines and carry out A line trials again in the following year (Fig. 9).

The selection of elite plants from A lines is continued until no further segregation occurs within the A lines, which takes two to four generations. Only then will a B line trial follow, and the breeding process is continued in the same manner as has been described for the selection of landraces. If selection begins in the F_2, breeding to constancy is rather long and drawn out.

The difficulties in selection arising from the strong heterozygosity and diversity of forms following variety crosses can be overcome by initial field propagation of the cross product starting from the F_3 on, and beginning selection within the population only in the F_5 or even later generations. Since in every population of self-fertilizing crops the homozygous plants will only have homozygous offspring and the heterozygotes will always segregate homozygous forms, an increase of homozygotes and a decrease of heterozygotes must take

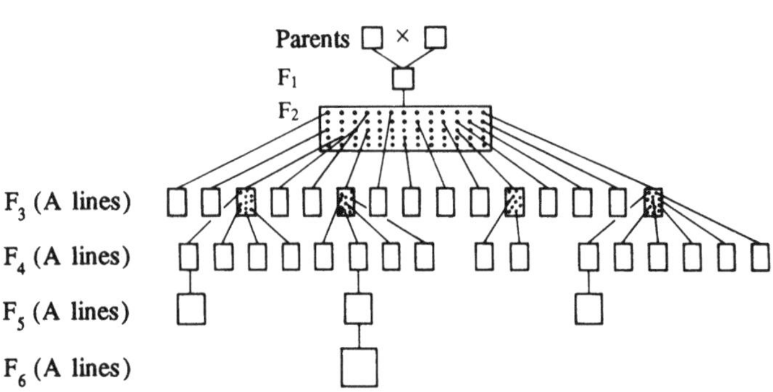

Fig. 9. Combination breeding starting in F_2

place over the course of generations. The portion of homozygous forms can be calculated for each generation with freely segregating genes as follows:

$$((2^m-1)/2^m)^n.$$

In this formula, m stands for the generation and n for the number of genes. If, for example, three genes are assumed (AaBbCc), a portion of homozygotes amounting to 0.77 = 77% can be expected in the F_4 generation. These propagations lead at the same time to a natural selection, so that all genotypes not adapted to the natural conditions are gradually eliminated. This method, which is called bulk breeding, is often applied in practice. Its advantage lies in the fact that the breeder is provided with already preselected and largely homozygous material, if he begins selection in the F_5 or even later generations (Fig. 10). Among the progeny of the selected plants, the number of segregations and negative variants will be considerably smaller than if selection had already started in the F_2. For successful practical implementation of bulk breeding, it is essential that the population be cultivated on as large a scale as possible and be grown under conditions appropriate for the variety to be bred. This is the only way to realize all gene combinations with an assured probability, and only with this method will natural selection also proceed in the direction intended by the breeder. The amount of time needed for bulk breeding can be decreased by application of the so-called single-seed descent method. For this, only one seed is chosen at random

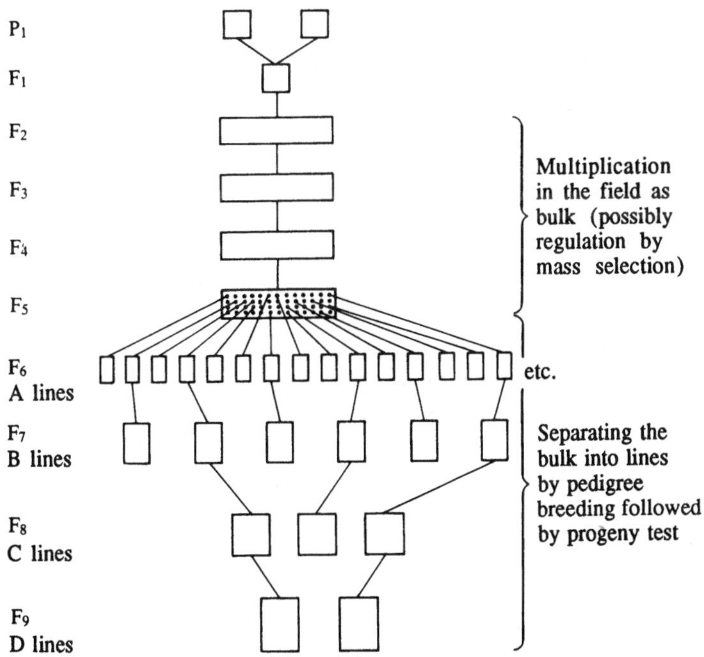

Fig. 10. Combination breeding by application of the bulk population method

from each plant for further use. Since the number of plants to be cultivated is considerably reduced by such a procedure, the individual generations can be cultivated in greenhouses, where two to three generations per year can be achieved.

In each bulk breeding, it is absolutely necessary that the breeder pay utmost attention to the natural selection processes within the bulks, in order to prevent natural selection from working against the desired goal. For example, in cereals, plants producing many small grains will prevail much quicker than those producing only few, but larger, grains. Thus, nature breeds varieties with low 1000-kernel weight, whereas the breeder often desires those with a higher one. By simply passing the seed through a sieve after each harvest, however, it is possible to exclude the small grains from further propagation and, by such a selection, to balance out the unwanted consequences of natural breeding selection. Also, the breeder's observation that different genotypes all develop equally well when cultivated in monoculture, but that in mixed cultivation one genotype is completely suppressed by another due to competition, is of importance for the right direction of natural selection in bulks.

Thus, it was ascertained in a bulk multiplication of crosses between glumed and naked barley that only few plants with naked grains are still present when selection takes place in the F_6 (recessive character). Therefore, in competition, naked grains are inferior to glumed grains[2]. In spite of this, the genotype that is inferior in competition may be as valuable as the successful genotype when there is no competition. The possibility of influencing the strength and direction of natural selection in populations by regulating exterior conditions can largely be taken advantage of by the breeder. If, for instance, cereal varieties with the ability to utilize maximum nitrogen dosages efficiently are to be developed, the breeder must cultivate the cross mixture starting from F_2 with especially high nitrogen dosages, so as to promote the natural selection of plants that react to such nitrogen applications with high seed production. In this way, the selection by the breeder can largely be supported and facilitated by natural selection. Populations of crosses in which the aim is to combine resistance of a race to animal or plant pests with economically valuable characters of another race must be cultivated under conditions especially favorable for the respective pest, in order to eliminate all heritable susceptible combinations.

Undesired natural selection can also be countered by other appropriate methods. In the above-mentioned crosses between glume × naked barley, it is advisable to select the naked grains already from the yield of the F_2 plants and to further cultivate them as separate bulk, in order to withdraw them from competition with glumed forms. Since the trait of naked grain is recessive, the isolated

---

[2]In glumed barley, the glumes grow together with the tegument; in naked barley this does not happen. During threshing, the naked grain is released from the glume, as in wheat. The reasons for the inferiority of the naked barley may, among others, be due to damages from threshing the naked grains and stronger susceptibility of the germinating seeds in the ground, and also because of photosynthesis deficiency of the glumes.

new bulk from plants with naked grain remains constant. Natural selection will now only be effective within the naked-grained plants and, over the course of generations, will lead to an increase in useful naked barley forms; these would have been inferior to the glumed ones in competition and therefore eliminated. The remaining bulk with the glumed barley will repeatedly segregate naked forms in subsequent generations as long as there are still plants within the glumed barley that are heterozygous for this character. If the naked grains are selected from each generation and separately propagated, the numerical proportion of segregating recessive naked barley plants will, of course, become less from year to year.

The opinions differ as to whether selection starting from F_2 or application of the bulk method is more successful. Comparative investigations have led to different results. The success of combination breeding not only depends on the method applied, but also on the respective crosses and their segregations, as well as on the skill of the person working with the material. Opponents of the bulk method point out that the greatest genetic variation with a majority of heterozygous types is only present in the F_2, and that this is too restricted in a perennial propagation, to the disadvantage of subsequent selection. Of course, this is especially true if no attention is paid to the natural selection processes during bulk propagation or when natural selection is guided in an undesired direction.

A compromise between the two selection methods described thus far is the bulk progeny method, which was first tested on cotton (Fig. 11). A large number of good phenotypes is selected from the F_2 and separately cultivated further as A lines (= families) until the F_5 or F_6. Then, as soon as enough seed is available, they are entered into a performance trial along with some standard varieties. During this time period, no individual plants are selected; however, families that are particularly inferior on the strength of the performance trials are completely excluded. After the F_6 generation, the most efficient families are chosen, based on the results of all trial years, to be bred to constancy by new individual selection followed by progeny trials. The majority of the selected plants will already prove to be homozygous. This breeding method has the great advantage that, on the one hand, it takes advantage of the automatic increase in homozygosity over the course of generations and yet simultaneously preserves the stronger heterozygosity of the original material.

Lately, so-called composite crosses have been developed to increase the chance of formation of new combinations. Composite crosses are bulk progenies composed of many different F_1 progenies. If, in addition, male-sterile genotypes (plants not producing any pollen) are used, the number of new combinations within a bulk can increase, since the free pollen grains that also occur to a certain extent in self-fertilizing crops may be able to fertilize.

For the formation of male-sterile plants, several new strategies are in use. The application of chemicals to reduce pollen formation, so-called gametocides, could not fulfill breeders' expectations. The incorporation of a foreign gene, an RNase coding gene coupled to a pollen-specific promotor, which causes such a high level of RNase in the pollen grain that it deteriorates, is presently undergoing practical testing.

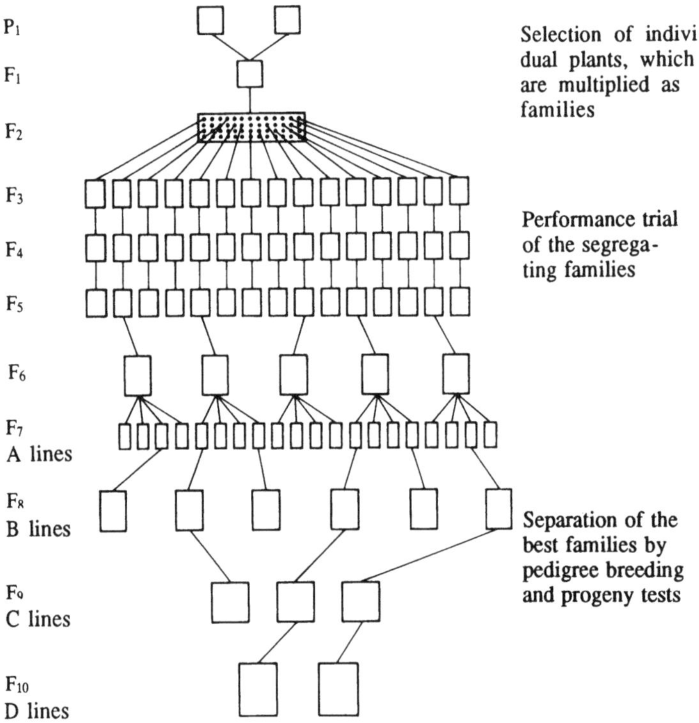

Fig. 11. Combination breeding using pedigree breeding followed by progeny tests (partial bulk)

A very effective instrument in combination breeding is the backcross method. It consists of another cross between the F_1 hybrid or a genotype selected from the F_2 and one of the parent plants. Such a backcross is repeated several times in subsequent generations, if the task is to transfer *one* character of a variety to another so that its general character as such is maintained, and if the variety is only to be improved in one trait, e.g., resistance to a parasitic fungal disease. The genetic variation of the backcross progeny compared to the F_2 generation is initially reduced by the backcross of a hybrid with one of the parents. If an F_2 is cultivated from a monohybrid plant Ii, three genotypes – II, Ii, and ii – are formed; if, on the other hand, hybrid Ii is backcrossed with the recessive parent ii, only the genotypes Ii and ii are formed. If it is assumed that both parents P_A and P_B differ from one another in many genes, an accumulation of the genes of P_B takes place after each backcross between the hybrid $P_A \times P_B$ with P_B, whereas the genes of P_A are constantly displaced. Purposeful selection of the desired character of P_A will again lead to the formation of the original P_B parent, enhanced with the valuable trait of P_A. The backcross parent is termed "recurrent" parent and the other parent, which only contributes *one* character to the cross product, is termed "donor" parent.

Selection within the backcross generations is very simple if the character to be transferred from the donor parent is dominant. The cross with the recessive recurrent parent always leads to dominant heterozygous plants with the desired character, which can be used immediately for another backcross with the recurrent parent. If, on the other hand, the character to be transferred is recessive, the heterozygotes suitable for a backcross cannot be distinguished from the dominant homozygotes after the backcross with the dominant recurrent parent. It is, therefore, advisable to let the desired recessives segregate again to cultivate separate self progenies from several backcross plants.

The advantage of the backcross method is that only a few hundred plants are needed to achieve breeding success within the entire breeding cycle, whereas in selection starting from F_2 and subsequent generations, a large number of plants must be cultivated and, above all, a large number of performance trials of the progenies must be carried out. Since in the backcross method the recurrent parent is fully restored again in its original characters -- improved by one character -- the performance trials can be omitted. Another advantage derived from this is that several generations (two to three) can be cultivated in 1 year in the greenhouse, i.e., the breeding cycle is accelerated and breeding costs are reduced. The increased labor connected with the artificial crosses in some cultivated species is unimportant when compared to the advantages of the backcross method.

A special form of the backcross method is convergence breeding. It is particularly applicable if two quantitative characters of the parents controlled by polygenes are to be combined. In the F_2 of such a cross, a large continuous variation occurs, because of which it is especially difficult to identify the desired combinations, since the phenotypic expression of polygenically determined genotypes can very easily be modified by environmental factors. In order to limit the undesired variation and simultaneously accumulate the desired combination types, two separate series of backcrosses are initially carried out with both parents P_A and P_B, which can very clearly be demonstrated by the following example drawn from tomato breeding.

The high ascorbic acid content of 47.20 ± 1.69 mg/100 g of parent P_A, which has a low fruit weight of 34.00 ± 1.8 g, is to be combined with the high fruit weight of 73.40 ± 3.6 g of parent P_B which, however, has only a low ascorbic acid content of 28.42 ± 0.80 mg/100 g. The series of backcrosses is carried out as follows:

1. Backcross ($P_A \times P_B$) × P_B
2. Backcross ($P_A \times P_B1$) × P_B
3. Backcross ($P_A \times P_B2$) × P_B

With each backcross the genes of the P_B parent automatically accumulate, i.e., the high fruit weight, while the breeder tries to maintain the original high ascorbic acid content of the P_A by selection.

In the other series of backcrosses, the process is reversed:

1. Backcross ($P_A \times P_B$) × P_A
2. Backcross ($P_A1 \times P_B$) × P_A
3. Backcross ($P_A2 \times P_B$) × P_A

In this series an automatic accumulation of the genes for ascorbic acid content takes place and the breeder must only select for large fruit.

As a result of these two series of backcrosses, the two original parents, now called "recovered lines" P'_A and P'_B, are mostly reproduced, improved in the genes for high ascorbic acid content and high fruit weight, that is, P'_A and P'_B have genetically converged. If an F_2 is cultivated from the cross between the genetically converged parents, the segregation will show less variation than in the F_2 of the cross $P_A \times P_B$. Indeed, the desired combination could be selected, namely, a line with 39.31 ± 0.98 mg/100 g ascorbic acid and a fruit weight of 69.8 ± 2.7 g. This aim was quite easily reached by means of convergence breeding.

The use of haploids, mentioned later, offers an additional means to characterize parents. By making 10–50 doubled haploids from an F_1 hybrid, it is possible to determine from the characteristics of the doubled haploid A_1 populations, whether, and which of the parents inherit valuable traits. This applies not only to self-fertilizing crops but also to outbreeders.

2.2.2.2 *Combination Breeding of Cross-Pollinated Crops*

In order to apply combination breeding in cross-pollinated crops, one must take into consideration that the mating system of allogamous crops is quite different. Consequently, populations of outbreeding crops display different behavior from generation to generation than do self-fertilizing crops. Due to self-sterility and inbreeding depression, it is difficult to develop homozygous lines in allogamous, cultivated species such as rye, sugar beet, carrot, and others. In this group of cultivated crops, one can proceed by carrying out a larger number of crosses using one individual each of two different varieties or origins in pairs. Since the cross parents are more or less heterozygous, the F_1 progeny of the pair crosses will already segregate, so that desired new combinations may already occur within this generation. The F_1 segregations of the pair crosses can be quite different from one another, depending on the different heterozygous compositions of the parent plants. To achieve a maximum amount of new genetic variation from crosses in cross-pollinated species, as many pair crosses as possible between two varieties are to be carried out. Selection can already begin within the best F_1 progenies of the pair crosses. The inferior offspring should be eliminated at once. The remaining F_1 progenies can be combined to form one population for selection in later generations, in order to simplify the handling of the breeding material. Thus, new panmictic populations are developed from which, with the help of pedigree breeding using progeny tests and simultaneous regulation of cross-pollination, an accumulation of desired gene combinations can be achieved.

For cross-pollinated crops, bulk breeding is not as important as for self-fertilizing crops, since here the laws of Hardy and Weinberg are effective. These state that under certain conditions the frequencies of the genotypes do not change from generation to generation. A population of outbreeders as a whole forms gametes with alleles A and a, the frequencies of which are termed p and q (= gene frequencies) and $p + q = 1$. Since these frequencies apply to the pollen grains as

Table 6. Frequencies of genotypes after random union of A and a alleles

♀\♂	pA	qa
pA	p^2 AA	pq Aa
qa	pq Aa	q^2 aa

well as to the egg cells, the genotypic composition of the next generation can be obtained from the simple chessboard pattern (Table 6).

Genotypes AA, Aa, and aa are to be expected in the numerical ratios p^2, 2pq, and q_2. If the frequencies of the gametes formed by this progeny are determined, the frequencies of A and a turn out to be p and q again. [The portion of the A gametes is determined by $(p^2 + pq)/(p + q)$; the portion of the a gametes amounts to $(q^2 + pq)/(p + q)$. By transformation, the formulae $p(p + q)/(p + q)$ or $q(p + q)/(p + q)$ are obtained, showing that the allele frequencies are p and q again.] From this it can be concluded that a cross-pollinated crop does not change with regard to its genotypic composition and, therefore, no accumulation of homozygotes can be expected if the bulk breeding method were applied.

Nevertheless, the constancy of the genotypic composition of subsequent generations is not always simply given. For a population consisting of 25 AA, 10 Aa, and 65 aa genotypes, a p of 30/100 = 0.3 and a q of 1 – 0.3 = 0.7 can be determined. These allele frequencies, however, result in a subsequent generation having a genotypic composition that is totally different: $p^2 = 0.09$ AA : 2 pq = 0.42 Aa : $q^2 = 0.48$ aa, converted to 100 individuals (= amount of original generation) = 9 AA: 42 Aa : 48 aa. If the allele frequencies of this population are once again calculated, p = 0.3 and q = 0.7 again, so that in fact no further change of genotypic frequencies is to be expected for a further generation. This condition of a population is termed the Hardy-Weinberg equilibrium. If only one gene is investigated, a population that is not in equilibrium is again in equilibrium with regard to this gene already after one generation. This law applies to every single gene locus. If, however, several gene loci are considered at the same time, adjustment of equilibrium takes correspondingly longer for the genotypes in question. Once the equilibrium is established, the genotypic frequencies of all subsequent generations remain constant.

In order for the Hardy-Weinberg law to be valid in this form, quite a few requirements must be met. Initially, it must be guaranteed that each individual has an equal chance to pair with every other individual of the population, which will only be the case with an optimum mixture of genotypes and fertilizing pollen (= panmixis). Thus, the Hardy-Weinberg law can also be applied to dioecious species such as hemp, spinach, and others (Table 7). Of course, no changes in allele frequencies, e.g., by mutations or by constant introduction of gametes (pollen) or genotypes (seeds) from other populations (= migration), may occur. If this does take place, the composition of the following generations can be estimated on the basis of the Hardy-Weinberg law, assuming that the mutation or migration rate is known. Other modifications of this law must be applied if polyploid species or inheritance depending on the sex are involved (Fig. 12).

Table 7. Derivation of the Hardy-Weinberg law concerning dioecious organisms

	Frequency of genotypes: p2 AA : 2 pq Aa : q^2 aa			
Possible combinations	Frequency of combinations	Resulting frequencies of genotypes AA	Aa	aa
AA × AA	p^4	p^4		
AA × Aa	$4p^3q$	$2p^3q$	$2p^3q$	
Aa × Aa	$4p^2q^2$	p^2q^2	$2p^2q^2$	$2p^2q^2$
AA × aa	$2p^2q^2$	$2p^2q^2$		
Aa × aa	$4pq^3$		$2pq^3$	$2pq^3$
aa × aa	q^4			q^4
Σ =	1	p^2 AA	2 pq Aa	q^2 aa

The size of the population also plays an important role. Strictly speaking, the Hardy-Weinberg law is only valid for infinitely large populations. Deviations from the law can only be observed in practice, however, if the population consists of just a few plants. The progeny of a 1 AA : 2 Aa : 1 aa population in equilibrium may be assumed to consist of just one plant. The probability that this plant has the genetic constitution AA or aa is 25% each, that is, with this probability an allele (a or A) will be lost completely. If two plants are cultivated, the probability for the loss of an allele will naturally be smaller. Four progenies of the population mentioned above will have the exact ratio of 1:2:1 at a probability of 18.75%. Chance alone will lead to the observation of other genotypic frequencies. In many of these cases, however, the allele frequencies will then be altered. These random, undirected changes, the unwelcome effects of which can be observed in small crops, are called "genetic drift." The smaller the population, the greater the danger that the genetic drift will lead to the loss of alleles within the course of one or more generations.

Finally, one must still point out of the type of selection that has a totally different effect on cross-pollinated populations than on self-fertilizing crops. Assuming that a self-fertilizing species consists of homozygous genotypes only, but still contains 10% undesired aa types, these could be removed by a single selection step. Since natural selection would, for example, only eliminate about half of the aa types each time, the decrease of these genotypes from generation to generation would take the following course if the breeder did not intervene: 10.0, 5.0, 2.5, 1.25%, and so on.

In cross-pollinated crops, the alteration of the genotypic frequencies after selection can be determined with the aid of the Hardy-Weinberg law. If, as in the above-mentioned example, the aa types are partially or completely removed before flowering, the allele frequencies of p and q necessarily change. In this case (selection of aa genotypes), the decrease in q from one generation to the next takes place according to the following formula: $-sq^2(1-q)/(1-sq^2)$ (Table 8). s stands for the portion of the selected plants. If the breeder makes an effort to

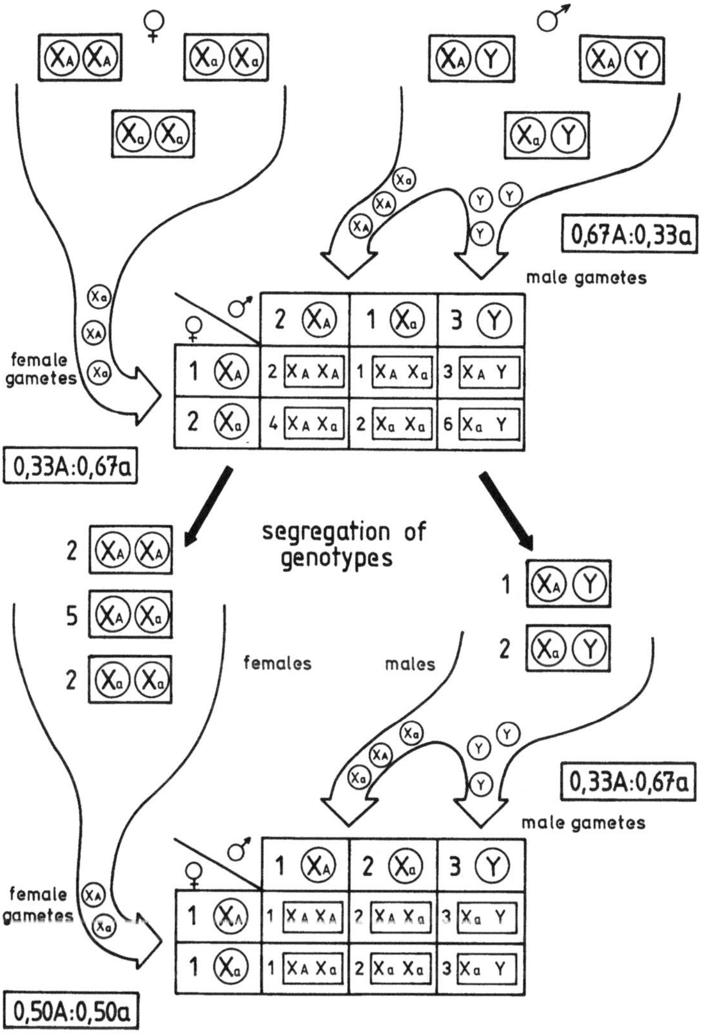

Fig. 12. Proceeding towards the Hardy-Weinberg equilibrium of one gene (A/a) linked with the sex chromosome (X)

eliminate all aa types within each generation as in the example of self-pollinating crops mentioned above, he must nevertheless expect the occurrence of the undesired forms: original population = 10.0%, the proportions of the homozygous recessives in the subsequent generations amount to 5.8, 3.8, 2.6, 1.9%, etc. ($q = \sqrt{q^2} = 0.3162$). The difference (Δq) between the old and the new q is calculated by $\Delta q = -1.0 \times 0.10\,(1 - 0.3162)/(1 - 1.0 \times 0.10)$. From this, a Δq of 0.0760 results. Consequently, the new q is: $q = 0.3162 - 0.0760 = 0.2402$. By squaring this value, the portion of aa types in the first subsequent generation is

Table 8. Change of allele frequency by one generation of selection against a recessive gene (After Falconer 1981)

Initial frequencies of genotypes			New frequencies for the allele a	Change of allele frequency
AA	Aa	aa		
p^2	$2pq$	q^2	q_1	$q = q_1 - q$
1	1	$1-s$	$\frac{q-sq^2}{1-sq^2}$	$-\frac{sq^2(1-q)}{1-sq^2}$

obtained: 0.0577 or 5.8%, and so on. Assuming a natural selection rate of 50% each time for comparison with the example for self-fertilizing crops (see above), the following series of the decrease in aa types is obtained: 10.0, 7.9, 6.3, 5.1, 4.2%, etc.

Naturally these laws also apply if, as in combination breeding, two or more characters controlled by different genes are to be combined in a way planned by the breeder. The different behavior of self- and cross-pollinated crops regarding the selection of desired combinations or the elimination of undesired ones is of considerable significance for success.

2.2.3 Haploids in Combination Breeding

Combination breeding, described in the previous section, results in a steady improvement of new cultivars. This improvement can be measured in an average yield increase of 1–2% per year. Although such classical approaches promise continuous breeding progress, these procedures are not very flexible. The solution of a new breeding problem, e.g., a new disease, takes time. On the average, the incorporation of a new character takes 15–20 years from the first cross until a protected cultivar is produced. The reason for this is that the growers must be provided with uniform material, at least in all important phenotypic and quality characteristics. In self-fertilizing crops this means a cultivar only has a chance on the market as a completely homozygous type. To achieve variety protection, homogeneity of outbreeders and vegetatively propagated crops or homozygosity of inbreeders is a prerequisite. Breeding for homozygosity is particularly time-consuming for traits with a dominant inheritance, as an AA genotype cannot be phenotypically distinguished from the Aa type. Techniques were highly desired to speed up this process.

Such suggestions were made by Blakeslee and Belling (1924), soon after Bergner discovered the first haploid higher plant in 1920, a plant with a simple set of chromosomes (1n). Melchers and Labib continued this discussion and added practical breeding aspects after new methods were developed in the 1970s, with which sufficient numbers of haploids could be produced. In the meantime, the technique has been improved to such an extent that in crops like potato, rapeseed, rice, barley, and wheat, haploids have become an integral part of commercial breeding programs. In barley the first variety, "Mingo," of haploid descent (by the *Hordeum bulbosum* procedure, Table 17) was licensed in Canada in 1980; in wheat the first variety, "Florin," based on the use of androgenetic haploids, was licensed in France in 1985.

As described in Section 3.4.3 in more detail, haploids may have one of two origins: either from the unfertilized egg cell (a special case of parthenogenesis) or from young pollen grains (microspore androgenesis). For breeding purposes, androgenesis in particular has become important since, in most crops, functional plants can be regenerated by culturing microspores. Altogether, the regeneration of microspores today has been used to induce haploids from more than 100 species of 50 genera. They are normally cultured within the anther, which is why this type of haploid production is often called anther culture. Haploids are completely sterile, smaller than diploids, and slow growing. During regeneration of plants from structures of haploid origin, a high percentage spontaneously becomes diploid or their genome may be doubled by colchicine treatment. Completely homozygous plants are the result of such doubling procedures. Figure 13 shows the most striking advantage of this process. From a heterozygous plant Aa, one obtains, after meiosis, a population of gametes in a 1A : 1a ratio. Regeneration of such gametes and doubling the haploid genomes result in a population of homozygous AA and aa plants in the same 1:1 ratio. In a classically produced F_2 from a heterozygous mother the proportion of, e.g., homozygous recessives is just one-fourth because of a 1:2:1 segregation.

Table 9 documents how drastic the increase in population size depends on the genome level when specific traits are desired. If, for example, the important characters are controlled by recessive genes, this means that for five alleles, the homozygous state shows up once among 1024 individuals in a diploid organism, in a tetraploid, once among 60 million plants, but in a haploid, once among 32 plants. Thus, the use of haploids not only speeds up the production of homozygotes but also allows work to be carried out with smaller populations or, if the same population size is used, permits the detection of relatively more and better genotypes. An even more important point is the fact that quantitatively inherited characters have also been fixed homozygously and can be used with high heritability in further crosses.

Even the most successful culture procedure does not result in more than 10% of microspores regenerating into plants. This might imply some type of selection. On the one hand, genomes that contain lethal but normally recessive alleles will not grow, resulting in a purification of such sublethal alleles (microspore sieve); on the other hand, there is possibly a selection for genotypes that allows regeneration of such single cells. No proof is available today that by this type of selection for regeneration capacity, genes that are negatively linked to low regeneration capacity are lost. With increasing numbers of molecular markers, particularly RFLPs, such questions will soon be answered.

While the experimental details for haploid induction will be given in Section 3.4, some principles on the use of haploids in combination will already be described here.

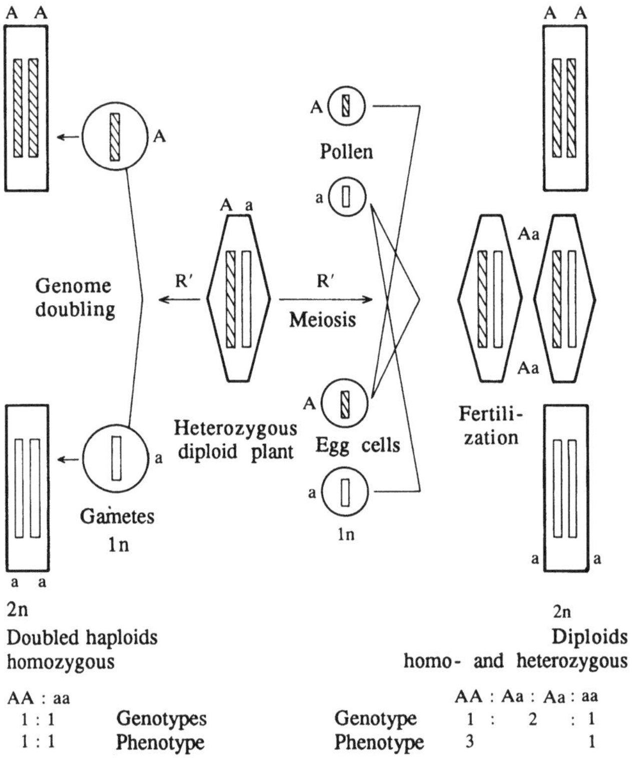

Fig. 13. Different segregations of a heterozygous genotype after the sexual (right) and the doubled haploid procedure (left)

Table 9. Ratio of completely recessive genotypes in F_2 for doubled haploids, diploids, and tetraploids

	One of (×) plants is recessive in		
Number of alleles	Doubled haploids	Diploids	Tetraploids
1	2	4	36
2	4	16	1296
•			
•			
5	32	1024	60 466 176
•			
•			
n	2^n	2^{2n}	6^{2n}

For applied purposes, it does not matter whether the doubled haploids (DH) are produced via parthenogenetic or androgenetic procedures. The cheapest and most effective technique, which results in 50-100 haploids from one donor genotype, will be used. In most cases today this is the anther culture procedure, not only because of its handiness but also because a plant normally produces more microspores than megaspores. Consequently, a higher number of individuals can be screened on the male side. Depending on the nature of the starting material as well as on the breeding aims, one or more haploid steps are recommended. F_1 hybrids or selfed progenies in later generations may serve as starting material for the haploidization. It has been demonstrated that one haploid step followed by selection in the greenhouse and the field during the first androgenetic and two subsequent selfed generations is the most effective procedure, if characters from related varieties are to be combined. For breeding programs that include wild types or unrelated genotypes, recurrent selection alternating with several haploid steps is very efficient.

To obtain homozygosity in an inbreeder six selfings are the rule. Each selfing is a sexual event with recombinations that will sometimes separate more alleles than desired. Furthermore, the selection for quantitatively inherited traits in heterozygous generations is often impossible, as all recessive traits are not yet visible. In doubled haploids the final form is already reached, allowing a very secure selection also for such quantitative traits. The selected individuals may be crossed or backcrossed already in the first androgenetic generation (A_1). The final step of 3–4 years bulk breeding is obtained immediately. This allows a rapid crossing sequence and, if necessary, further haploid steps. The degree of relationship of the two parents determines the number of backcrosses needed. The recurrent selection alternating with haploid steps leads to new homozygous lines, which can be tested for agronomic value in replicated field experiments. If a positive selection is possible in the first doubled haploid (DH) plant population, e.g., with gene probes for definite characters, the backcross can be done in this generation. The earlier an effective selection can be obtained, the shorter is the breeding time. If the selection can be done on a single plant without destroying this plant, then it will be possible to shorten the recurrent selection alternating with haploid steps in the way shown in Fig. 14. This will be true for most inbreeders since the rates for haploid induction are still increasing. The use of haploids is probably the best procedure available today for combining quantitatively inherited characters of sexually propagated crops. (For such combinations in vegetatively propagated crops, see Sect. 3.4.3.)

Three examples for the commercial use of haploids in breeding programs follow.

1. Barley. Of particular interest to barley breeders was the rapid incorporation of resistance to the soil-borne barley yellow mosaic virus (BaYMV). The use of doubled haploids offered a rapid approach for incorporating the BaYMV resistance into commercial cultivars. More than 6000 androgenetic lines were produced from microspores of F_1 anther donor hybrids of crosses between susceptible, but high yielding and resistant, yet less productive cultivars. Since 1984, increasing numbers of homozygous doubled haploid lines carrying the

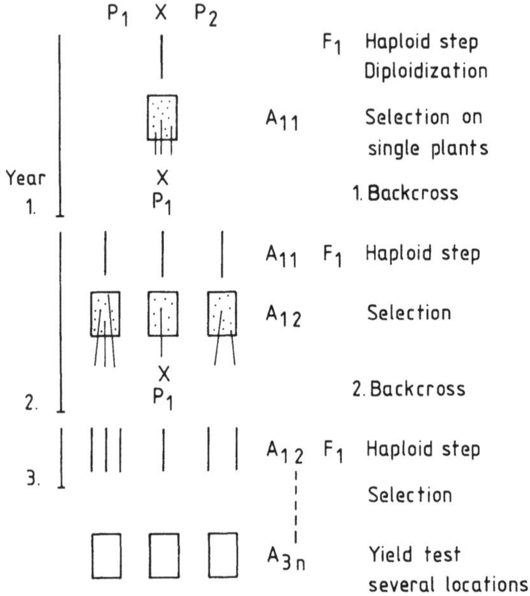

Fig. 14. Scheme for recurrent selection alternating with haploid steps using an early selection system in the A_1, e.g., molecular markers

BaYMV resistance were released to practical breeding companies (Table 10). Several such lines are undergoing official testing by the German Seed Board.

2. Rapeseed. In *Brassica napus*, mainly quality characteristics are emphasized in haploid breeding techniques. Here vigor of microspores was positively correlated with high glucosinolate content in the androgenetic regenerants. Very low or very high glucosinolate levels are various qualities needed either for human consumption (low) or for industry. The spontaneous selection here favored only the nonfood. Today, particularly in *Brassica*, the isolated microspore culture (culture of microspores in liquid culture media in isolation from the anther) is successful. Isolated microspores apparently offer a more

Table 10. Breeding for barley yellow mosaic virus (BaYMV) resistant winter barley lines via androgenetic doubled haploids (Foroughi-Wehr unpublished)

Year	Material	Numbers	Type
1983–85	P1 (S) × P2 (r) P1 (S) × P2 (r) × Igri (S)	59	Parents
1984–86	F_1	38	Anther donor hybrids
1984–86		666,810	Anthers plated
1984–86	A_1	6,626 (1%)	Doubled haploid lines
1985–88		3,907	Lines tested for BaYMV
		1,849	Lines (S) susceptible
		2,058	Lines (r) resistant

efficient system of regenerating a random sample from the microspore population than does anther culture. This is particularly important when the desired trait – such as low glucosinolate content in *Brassica* – is linked with low plasticity, resulting in rather small regeneration rates of such genotypes.

3. Potato. In potato a stepwise reduction of the tetraploid ploidy level (2n = 2x = 48) offers the advantage of a simpler inheritance and also a better chance to combine qualitatively inherited characters. Data now available demonstrate that field resistance to potato virus Y (PVY) and potato leaf roll virus (PLRV) is maintained and expressed on all ploidy levels (1x, 2x, and 4x), which means that both monogenically and polygenically inherited resistances are maintained during the successive haploidization process. When a genome has passed the monohaploid level, a homozygous additive character can be transferred efficiently to the next generation via classical crossing. Similar results have been obtained for polygenic nematode resistance. In a vegetatively propagated crop such as potato, the F_1 can be propagated as a new cultivar (cf. Sect. 3.4.3).

In outbreeders such as rye, *Secale cereale*, there are still tremendous difficulties in using haploids. Due to the heterozygous nature of cultivars it is difficult to standardize the procedures, but even with inbred lines the production of doubled haploids in frequencies sufficient for application has not yet been achieved. Nonetheless, even if the amount of doubled haploids were sufficient, their further use in breeding programs would be limited, as most lines will be self-sterile. Similar problems exist in most legumes.

2.2.4 Limitations of Combination Breeding

Using combination breeding, the possibilities for the development of new varieties that are more efficient in both qualitative and quantitative respects are extraordinarily large; however, combination breeding is also limited in certain ways. The reasons for this limitation can first of all be of a purely genetic nature. The genes are located on the chromosomes. In a hybrid, the single chromosome sets of both parents are combined so that all chromosomes are present in pairs, but the maternal and patternal genes – or more precisely, the alleles of the gene loci – remain separate in association with one another in their respective chromosomes. Only in meiosis, the process of gamete formation, does a recombination of the male and female chromosomes take place by reciprocal exchange of corresponding chromatin fragments. This process of meiosis is associated with the recombination of partners of homologous chromosome pairs. Its individual phases are shown in Fig. 15. It is important that the paired chromosomes are split lengthwise and thus consist of four so-called chromatids. However, recombination only takes place between two nonsibling chromatids, i.e., between chromatids, of which one is of female and the other of male origin. At the end of meiosis a tetrad of four gametes is formed, two of which represent the original male and female chromosomes, whereas the other two are mosaics of male and female chromatids.

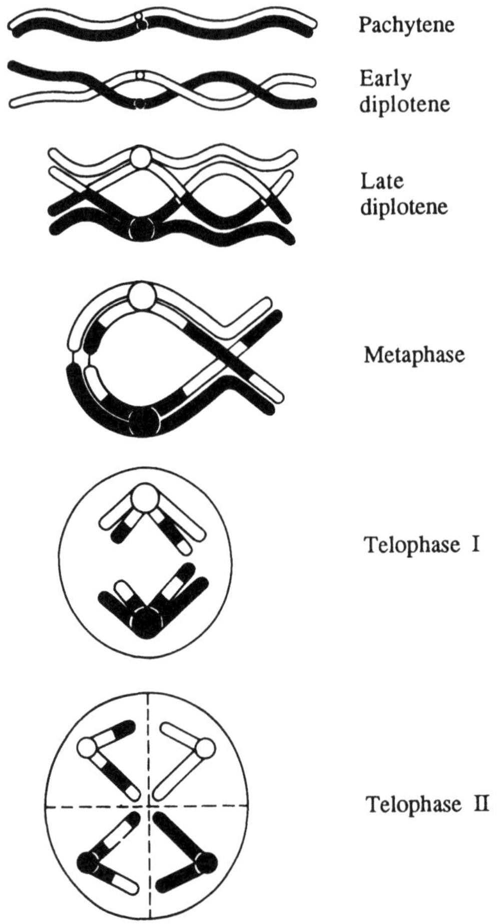

Fig. 15. Diagram of pairing and crossing-over processes. Recombination between male and female chromosomes seen in telephase (Straub, in Kappert and Rudorf 1958–1962)

The process of chromatid exchange between two nonsibling chromatids is characterized in the figure by the intersecting of the chromosomes, a so-called crossing-over. The overlapping points are called chiasmata; they are interpreted as being the cause or the consequence of a crossing-over.

If on one chromosome there are alleles A and b and on the corresponding homologous chromosome alleles a and B, the prospect of attaining the desired combination AB by crossing-over increases the farther apart the genes are on the chromosome. If they are situated very close to one another, crossing-over can very seldom be expected, so that an extensive F_2 with a large number of heterozygous plants of the constitution Ab/Ab must be used in order to accomplish this event. If the double heterozygotes have largely been replaced by the homozygotes Ab/Ab or aB/aB, respectively, in the following generations – the F_3 consisting of only 25% of double heterozygous plants – the breeder has no more possibility of selecting the desired combinations AB/AB or ab/ab. On the other hand, selection prospects applying the backcross method are much more favorable, since then the heterozygous form is produced anew in each generation, from which the desired combination can be formed by crossing-over.

If there is no crossing-over, a backcross between Ab/aB with an aB/aB always results in 1 Ab/aB: 1 aB/aB genotypes again, and with them, the possibility for the expected crossing-over remains. If an AB/ab genotype occurs within a backcross progeny of an Ab/aB genotype due to crossing-over, 25% of the desired homozygous AB/AB genotypes will segregate upon subsequent selfing.

A genetic mechanism, which can prevent a phenotypically appearing and constantly inherited new combination, is caused by multiple alleles. It is characterized by the presence of more than two alleles to one gene locus. Thus, three alleles of one gene locus have been described in the ornamental plant *Godetia*, which results in a varied coloration of the petals.

1. Petals without spot ff
2. Petals with central spot F^CF^C
3. Petals with basal spot F^BF^B

A cross of these three genotypes always leads to a monohybrid segregation, which is characteristic for the elements of a series of multiple alleles. It is interesting that the hybrid F^CF^B displays a basal as well as a central spot. This combination cannot be bred true, however, since it only occurs as a heterozygote.

Combination breeding can also be complicated by the pleiotropic effect of a gene, because several characters seem to be controlled by one single gene or by several tightly linked genes. This question will be discussed in more detail in Chapter 3, as well as the phenomenon that genes, especially those with pleiotropic impact, display altered expression of characters in a background changed by combination breeding.

Combination breeding has led to great progress in many cultivated species. Its effectiveness has by no means been exploited yet, and will not be as long as the sources of natural diversity of forms are not exhausted or further genetic variation is generated by the introduction of mutations. In this context, the effectiveness of the systematically performed backcross method with its many variations is again pointed out, which has often not been adequately noted and applied in practical breeding.

If combination breeding is first initiated in a cultivated species, one should proceed step by step, i.e., one should begin with a cross between two varieties or landraces that are largely adapted to the environment of their future habitat, with the aim of combining certain valuable characters. By crossing valuable, but different lines derived from this first cross again in the next step, further progress can be relatively quickly made, especially through the occurrence and use of transgressions.

Once the genetic potential of such a cross has been exhausted, the breeder should not hesitate to introduce genes or gene complexes of variants that come from another ecological environment. However, since especially in crosses between forms of different ecological adaptations, the combination products to be expected cannot be predicted, a large number of foreign genotypes with the desired character should be used for the cross with the variety to be improved; i.e., variety A should be crossed with varieties B, C, D, E, etc. Another step to obtain promising selection material is a further mutual cross between the F_1 hybrids (A × B) × (A × C), etc. If the desired character to be selected is determined differently in B and C, transgressions can be expected. If gene material of varieties that strongly differ ecologically is used, restoration of the gene complexes for adaptation to the corresponding environmental conditions will not be possible without a backcross with the original variety.

2.3 Hybrid Breeding

In the sections dealing with selection and combination breeding, methods were described for breeding uniform and stable varieties, i.e., varieties whose most important valuable characteristics are transferred to their offspring by generative propagation. These varieties contrast with the so-called hybrid varieties, also called F_1 hybrids in certain cases. These are of course also very uniform, however, they do not transfer the valuable characteristics in their entirety to their progeny, but segregate in the subsequent generation (F_2). On the other hand, if the combination partners are correctly chosen, an F_1 variety can be created that noticeably exceeds the efficiency of the original material. This hybrid vigor in particular is an incentive to many breeders to produce seed of F_1 varieties which, however, can only result from constant crosses of two parent lines. Only if the parent lines are largely homozygous, but genetically different from one another, can heterozygous, uniform hybrid variety be obtained that segregates in the following generations.

Since the character of the variety is lost again in the following generations due to genetic segregation, hybrid varieties offer the breeder a natural protection of his intellectual property. Another breeding advantage lies in the fact that often valuable characteristics of two parents can be combined more quickly and easily in a heterozygous F_1 hybrid than in a variety that must be bred to the stability of desired characters. This is true for polygenically as well as for monogenically inherited characters.

2.3.1 Hybrid Breeding of Cross-Pollinated Crops

The methods for breeding hybrid varieties in cross- and self-fertilizing crops differ in many ways. They are therefore dealt with in separate sections. The breeding of hybrid varieties began with cross-fertilized crops. Its great success in maize breeding caused similar methods to be developed for self-fertilizing species.

2.3.1.1 Occurrence of Inbreeding and Heterosis

In the discussion on selection methods for outbreeding crops, it was already pointed out that the most effective way to regulate cross-pollination is by forced self-fertilization. In some outbreeding crops, such as maize, rapeseed, celery, and cucumber, self-pollination succeeds easily. Other species are more or less self-sterile, so that selfing only results in little or no seed set.

In progenies of selfed plants, termed inbreeding or I generations, a reduction in growth can very often occur, which decreases with the number of I generations until an inbreeding minimum is reached. The I lines either stay the same or die due to lack of vitality and fertility. Furthermore, the segregation of lethal or sublethal forms, e.g., chlorophyll-deficient plants, dwarf types, or other monstrosities, can be observed in the first I generation (I_1). Since these generally occur at a frequency of 25%, they can each be attributed to the effect of a homozygous, recessive gene. In a cross-fertilized population, these recessive genes being heterozygous, are protected by their dominant allele. Only if two plants heterozygous in the same gene happen to pollinate each other – a relatively rare event – or if a heterozygous plant is by chance selfed by the breeder, does a new phenotype segregate in the progeny as a homozygous, recessive genotype at a frequency of 25%.

The progressive reduction of growth in the I generations can also largely be explained by the segregation of recessive polygenes, i.e., genes present in large numbers having only little significance for the phenotype, but which are cumulative in their effectiveness. They are also strongly modified by environmental conditions.

Likewise, the segregation of chromosomal mutations in a homozygous condition may lead to inbreeding depression. This includes the loss of small pieces (deficiencies) or the doubling (duplication) of chromosomes, which do not reduce the vitality of the individual in question in the heterozygous condition due to the compensating effect of the normal homologous chromosome.

It is doubtful whether all inbreeding phenomena can be attributed to the effect of Mendelian genes alone. Often a characteristic can already be completely uniform in the I_1 progeny, e.g., in onions with earliness and simultaneous ripening. The contribution of extranuclear (plasmatic) genes must also be considered. With selfing of cross-fertilized populations, not only can recessive deficiency characteristics segregate, but also recessive genotypes with economically important characteristics such as stability, earliness, resistance against disease, etc. Also, these valuable types obtained by inbreeding mostly display a more or less reduced vitality.

In all cases, inbreeding leads to a genetic differentiation of cross-fertilized crops and its usage for the breeder is therefore simplified and broadened.

By crossing it is possible to offset the depression due to inbreeding; some hybrids even surpass their original parent populations in growth and productivity, especially if I lines of different descent are used for the cross. They show a so-called heterosis effect, for which disappearance in the subsequent generations is characteristic. Inbreeding and heterosis seem to be two sides of one phenomenon. It is therefore obvious to explain both mainly by mutually complementary genetic mechanisms. This is based on the idea that the segregation and accumulation of homozygous, recessive genes in the I_1 generations are responsible for the occurrence of defective types and reduced vitality, and that heterosis is due to the stimulating effect of heterozygosity, which is realized to the highest degree in the hybrid of the cross of two I lines. This hypothesis of heterozygosity, which was independently stated by G.H. Shull and E.M. East as early as 1908, was later altered to the effect that, for heterosis, the heterozygous allelic condition of specific genes could be proven, meaning that heterozygosity in itself does not cause an increase in vitality. These specific genes are characterized by two further features: the homozygous recessive forms (aa) sometimes have a sublethal effect, whereas the heterozygotes (Aa) can be superior to the homozygotes (AA) as regards a performance character, e.g., chlorophyll production.

This so-called superdominance, also termed monohybrid heterosis, can be explained by the interaction of the dominant and recessive allele of a single gene locus. Heterosis due to superdominance can therefore not be bred to constancy. The same situation exists as in the series of multiple alleles, where the combined effect of two alleles can only be realized in the hybrid.

An explanation for the appearance of heterosis in the F_1 and the failure of trials for its fixation in subsequent generations is also given by the linkage hypothesis postulated by D.F. Jones in 1917. Jones based this hypothesis on the so-called dominance hypothesis, in which dominant alleles of genes have a stimulating effect, recessive ones, an inhibitory effect. Furthermore, he assumed a multitude of such promoting or inhibiting genes, which are distributed over all the chromosomes of an organism. As a consequence, linkages between dominant and recessive alleles of several gene loci positioned on the same chromosome are to be expected, as shown in the model of Fig. 16. The two inbreeding lines A and B each possess three dominant promoting genes in the homozygous condition, A, C, D or B, E, F, respectively, so that altogether there are six promoting genes in the heterozygous condition in the hybrid. In comparison to the homozygous parent there is a corresponding increase in the F_1 productivity. In the generations following the cross, homozygous combinations are formed again along with initial inbreeding, consisting again of fewer dominant genes due to the existing linkages, therefore leading to a decrease in growth. The possibility that chromosomes with the dominant genes A, B, C or D, E, F will be formed in the hybrid by breaking linkages and thereby providing the fixation of heterosis in the following generations should be very small. Tight linkages are seldom broken or even never if the chiasmata are limited to certain regions of a chromosome. It can also be imagined that increased growth can be brought about by the interaction of nonallelic genes.

It can be concluded that the characteristics of inbreeding and heterosis are controlled by a multitude of cytological and genetic mechanisms, and that the effects are much more complicated than could be demonstrated here in the few analyzed examples. These examples, however, have given some insight into the cytogenetic circumstances of inbreeding and heterosis and furthermore into the consequences to be considered regarding the breeding work of cross-pollinating

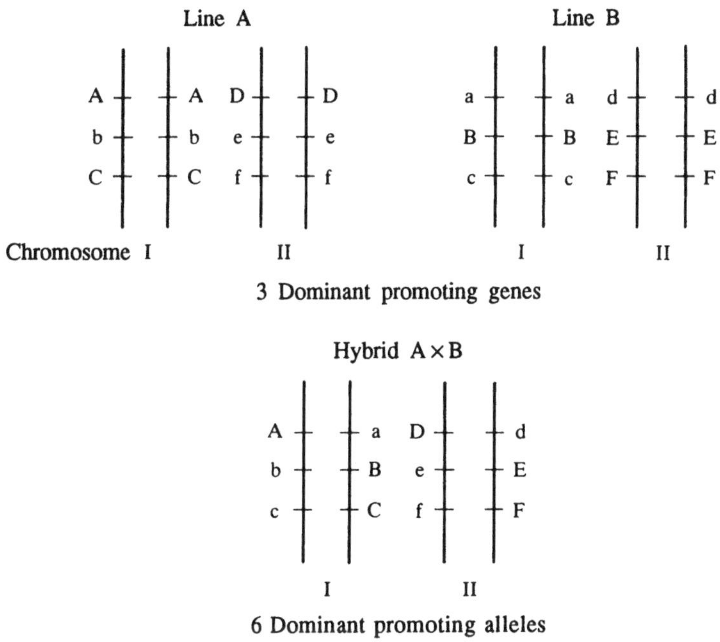

Fig. 16. Dominance hypothesis interpreting the effects of inbreeding and heterosis. Fixation of dominant alleles is prevented by linkage

crops, especially in the development of hybrid varieties. In natural populations of outbreeding species, a multitude of mechanisms are known that prevent self-fertilization and favor the permanency of heterozygous genotypes.

Heterozygous plants always display a certain level of heterosis. Only genotypes absolutely homozygous are free from heterosis. If two not completely homozygous inbreeding lines are crossed and if the cross product is inferior to its parents, it cannot be concluded that no heterosis at all has set in this case. Rather, the heterosis of the F_1 generation was smaller than that of the parents. Generally, the plant breeder will be interested in that part of heterosis which gives an advantage to the hybrid variety compared to the original material or other already existing varieties (Fig. 17).

Next, some biological specialties concerning flowers will be described, because they are of great importance for hybrid breeding. With appropriate usage, some of these actually make possible the production of F_1 seed in quantities usual in commerce. Most obviously, fertilization after self-pollination (autogamy) is prevented in plant species that are heteroecious, i.e., dioecious (Fig. 18). In these species purely male and female flowers are distributed to different individuals, so that hereby fertilization between two different plants is ensured in any case (spinach, asparagus, hemp, among others).

In monoecious species, female as well as male flower organs can be found on the same individual. In spite of this, autogamy is not readily possible in many

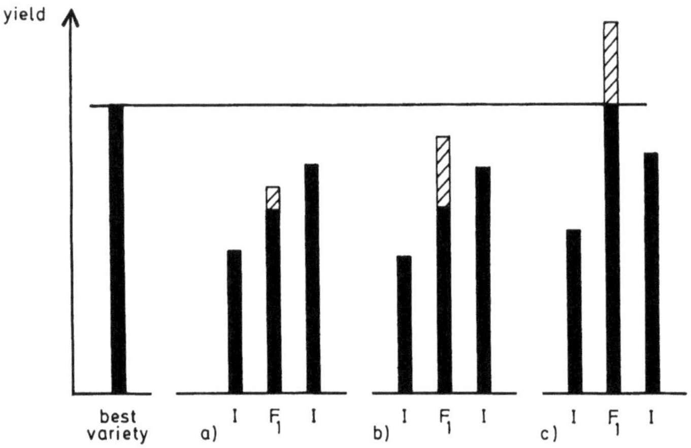

Fig. 17a-c. Different "types" of heterosis, *a* F_1 yielding more than the mean of the parents; *b* F_1 yielding more than the best yielding parent; *c* F_1 yielding more than the best yielding variety (*hatched areas*)

monoecious plant species. Thus, there are plants, the male and female flowers of which are separated by space (maize, cucumbers). These diclinous species are especially suitable for the production of hybrid seed, since the removal of the undesired male flowers of the female combination partner (mass emasculation) is, in principle, not difficult. This will be discussed in Section 2.3.1.4. If the female and male flower organs are combined in hermaphroditic flowers (bisexual flowers), self-fertilization can be prevented by a chronological differences in the ability to function of the two sexes; this is called protandry if ripe pollen is released from the anthers before the female organs are mature enough for fertilization. In protogyny, the circumstances are the opposite. In this way, self-fertilization can largely be prevented. Nevertheless, the protection against autogamy is not as pronounced, because under certain conditions neighboring flowers of the same plant could achieve pollination (geitonogamy). Also, morphological differences in the structure of the flowers may favor cross-pollination. In hercogamy, the spatial arrangement of anthers and stigmas counteracts self-pollination (pansies).

In addition to this, there is a number of plant species in which seed set after pollination with its own pollen is prevented due to a genetically conditioned, physiological mechanism. As a rule, inbreeding cannot occur in nature in such self-sterile (self-incompatible) plant species. Two kinds of self-incompatibility can be distinguished: heteromorphic and homomorphic. In heteromorphic self-incompatibility, there are different plant types depending on the length of the filaments and styles. Plants with long styles have short filaments and, vice versa, plants with short styles have long filaments. These morphological differences, together with a physiological barrier due to the sporophytic incompatibility, prevent germination of pollen on the stigma of the same plant and also mutual pollination of plants of the same morphological types.

In homomorphic self-incompatibility, a genetically conditioned incompatibility between the tissues of style or stigma and pollen grain of one plant prevents self-fertilization; it is due to physiological factors only. The heteromorphic as well as the homomorphic incompatibility mechanisms are used in some plant species for the production of hybrid seed. This will be discussed in detail in connection with hybrid breeding in Section 2.3.1.4. A schematic diagram of the biological relations of flowers that are of interest for hybrid breeding is given in Fig. 18.

Nature employs still other mechanisms to maintain heterozygosity, which should briefly be described. Thus, a system of balanced lethal genes exclusively leads to the formation of heterozygous genotypes. Lethal genes are characterized by the inability of homozygous recessive or dominant genotypes to reproduce and by the fact that the gene in question can only be passed on by the heterozygous plants. If two of these genes are linked, the heterozygous condition can be permanent without segregation of homozygous lethal combinations. In snapdragon, for instance, there is a gene Alb-alb (*albina*), wherein the recessive alb-alb genotypes produce no chlorophyll and therefore die in the seedling stage. This recessive lethal gene is linked to the dominant lethal gene Aur-aur (*aurea*). The aur-aur plants have green leaves, the Aur-aur plants yellow leaves, and the Aur-Aur plants are not viable as white seedlings. The alleles of these two genes are distributed on a pair of chromosomes as can be seen in Fig. 19.

If there is no crossing-over, only Aur-Alb and aur-alb gametes are formed. Only yellow plants result from this combination, which are heterozygous for both genes; the homozygous combinations are lethal (Table 11).

In certain *Oenothera* species, which are structural hybrids originating from two different species, the female and male genomes are each inherited as a

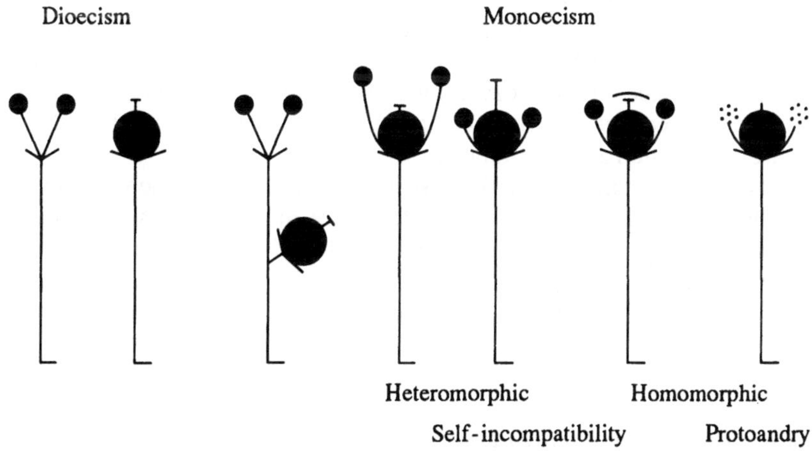

Fig. 18. Prevention of autogamy by different localization and function of the generative organs

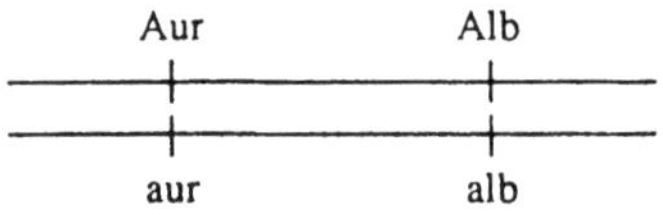

Fig. 19. Linkage of the lethal genes *Aur aur* and *Alb alb*

Table 11. Maintaining heterozygosity by two linked lethal genes

	Aur Alb		aur alb	
Aur Alb	Aur Alb Aur Alb	(lethal)	Aur Alb aur alb	
aur alb	aur alb Aur Alb		aur alb aur alb	(lethal)

complex; during meiosis, chromosomes of male and female origin migrate to the opposite poles as a coherent chain. In the further process of the formation of gametes and zygotes, gametic lethal genes prevent one of the complexes from developing as male gametophyte and the other complex as female gametophyte. Thus, only zygote combinations between two different complexes can be formed and the original structural hybrid is maintained. Instead of gametic lethal genes, zygotic lethal genes can also be effective by letting the zygotes containing two identical complexes die off.

As discussed in Section 3.3, seed formation takes place in apomictic forms without meiosis and consequently without recombination of the genes, so that the offspring has the same genetic constitution as the female parent. If this is heterozygous, then all offspring has the same heterozygous constitution.

The mechanisms briefly described here for prevention of self-fertilization and maintenance of heterozygosity indicate that in the plant kingdom, a heterozygous constitution must be of advantage to environmental conditions as regards adaptation and adaptability.

The obvious conclusion is that also in our cultivated crops heterozygosity can be advantageous for an increase in performance, at least in cross-fertilized crops. The highest degree of heterozygosity is reached if certain homozygous I lines, which have been developed from different genetic populations, are crossed. The obtained F_1 seed is put on the market as an F_1 hybrid, as was explained in Section 2.3.

2.3.1.2 *Production of Inbred Lines*

The first step in developing a hybrid variety is to produce inbred lines from suitable populations. When using this breeding method, the procurement of starting material with a large genetic variation is also of considerable importance.

The production of inbred lines by forced self-fertilization can easily be carried out with species that are largely self-fertile or with which a self-fertility brought about by environmental conditions (pseudo-fertility) can be used. If the self-pollination is done by enclosure of the flowers in bags made of gauze or parchment or other synthetic materials, the seed set is nearly always lower than by open-pollination, due to the changed micro-climate or plant disease. An improved seed set can be obtained if the pollination is done by insects, such as house flies in carrots and onions. If seed set fails after selfing self-sterile species, repeated pair crosses can lead to largely homozygous inbred lines. This method requires more generations than selfings, by means of which one to two generations may sometimes already lead to a degree of homozygosity sufficient for breeding purposes. Selection of self-fertile plants occurring as mutants in populations of cross-pollinated, cultivated species also offers a possibility to the breeder for overcoming difficulties in the production of inbred lines. Genetically, self-fertility can be due to an allele S_F of the S series or to a specific gene. For example, self-fertile genotypes are known in rye, in sugar beet, and in cabbage.

In the development of inbred lines, the selection between the I lines is of special importance. The degree of inbreeding depression is genetically determined, so that vital I lines can also be developed from one population. The greater the seed production of an I line, the more economical is its own propagation as well as the production of hybrid seed from crossing two I lines. If, in the development of I lines by repeated selfings in subsequent generations, particularly vigorous plants are selected for the next selfing, there is a risk of selecting mainly heterozygous plants and thereby delaying the desired homozygosity of a line.

One attempts to overcome the difficulties described here in the development of I lines by using principally the same methods that have already been specified for the production of pure lines in combination breeding. Selection or production of haploid plants followed by doubling of the genome are also quick and easy methods to produce homozygous diploid plants.

Using haploids in the hybrid breeding of cross-pollinating crops makes possible so-called "analytical breeding." This expression was coined for breeding processes in which cross parents (especially for the production of inbred lines) are preselected by making use of the single haploid genome. After such an analysis, the prediction and planning for the hybrid character is more reliable. As early as 1949 the first haploids were available in maize, and Chase (1963) already pointed out in the 1950s the possibility of using haploids for hybrid breeding in maize.

To be able to differentiate the occasionally appearing haploids from the large number of normal offspring plants in specific crossings, Chase employed a color marker gene that was homozygous dominant in the pollinator. All hybrids must then display the color marker, whereas the possibly parthenogenetically derived haploids (compare Sect. 3.4) should not. The frequency with which haploids were found in these early experiments depended particularly on the genotype of the female parent and, on average, amounted to 1 haploid plant among 900 seeds. A very high yield of haploids could be achieved by using parents that had already completed one haploid cycle. Kermicle (1969) found the mutant "indeterminate gametophyte" (ig), which raises the rate of haploids to 3% after transfer into any given genotype. These relatively high frequencies of haploids led to the discovery of androgenetic haploids (see Sect. 3.4). As in most plants, their cytoplasm originates from the female parent. Without involvement of the female nuclear genome, the male sperm nucleus develops without prior fertilization in the cytoplasm of the egg cell or the synergid. As explained under the special methods, this kind of androgenesis should be termed ovule androgenesis, in which the plasmon of the female parent can be combined with the genome of the male parent.

Besides the very rare ovule androgenesis, the generation of haploids in maize is successful today using microspore androgenesis (anther culture). Today, in 3–10% of maize anthers, the formation of embryos can be induced with common anther culture. In contrast to many other cereals, the majority of the regenerates (95%) is haploid and only a small group is spontaneously diploid. Furthermore, no albino plants occur, which strongly restrict practical use of androgenesis in some other cereal species. It is to be pointed out once again that during the haploid step a genome is purged of sublethal alleles. Therefore, it is not surprising that completely homozygous doubled haploids in rye or potato show good vigor and practically no inbreeding depression. Nevertheless, as was again found in rye and potato, after their hybridization distinct heterosis can be observed.

2.3.1.3 Testing for Combining Ability

In all probability no hybrids transgressing the performance of the original population will result from crossings of inbred lines with low intrinsic performance. If and to what degree heterosis can be expected after crossing two efficient I lines cannot be predicted from their performance per se. Therefore, in order to find the best combinations among a great number of I lines, testcrosses must be carried out in all possible combinations, the number of which can be determined by the formula $n(n-1)/2$. With 100 I lines, 4950 crosses, not considering the reciprocal ones, and a corresponding number of progeny tests would be necessary. In practical breeding such extensive experiments are not feasible. Therefore, by applying a less extensive method, preselection takes place by crossing these 100 lines with a "tester," i.e., instead of 4950 crosses, only 100 have to be carried out. After comparing the I lines × tester progenies by performance trials, it is determined which I lines have a so-called "general combining ability." The technical realization of these testcrosses, which are termed "topcrosses," depends on the biological manner of reproduction of the cultivated species.

In maize, a monoecious species, I lines and testers are planted in alternating rows. The I lines are then emasculated by removal of the male inflorescence at the tip of the plants, so that they are exclusively pollinated by tester pollen. Crossing plants with hermaphroditic (androgynous) flowers is more difficult. Here a tedious manual emasculation is often necessary. Only in self-sterile species can emasculation be omitted, if the cross is restricted to one individual plant or to a clone deriving from this plant, e.g., in gramineous species. The plants or clones to be tested are planted in a field crop of a tester, at a distance from one another that is sufficient to largely preclude mutual pollination by wind or insects (Fig. 20). It is recommended that the topcross be carried out at several locations and for several years. A more difficult problem with this method is the choice of a suitable test variety. After about 90% of the tested lines have been eliminated by the topcross, the specific combining ability of the remaining lines is determined by diallel crosses and progeny tests. For experimental setup and evaluation of incomplete diallel crosses, reference is made to Fyfe and Gilbert (1963).

As experience has shown, the general combining ability can already be determined with great certainty with I_0 and I_1 plants or lines, respectively, i.e., with largely heterozygous genotypes. This permits early testing.

In hybrid breeding the improvement of I lines as regards their combining ability and specific characteristics, e.g., resistance to diseases, is of special importance. For this, several methods are being used. "Gamete selection" consists of pollination of I lines with a variety or a population, with the goal of selecting gametes, i.e., carriers of genes, with which the combining ability of the I lines can be improved.

Compared to gamete selection, the methods of "recurrent selection" and "reciprocal recurrent selection" are of more importance in practical breeding. Using these two methods, populations with improved genotypic composition can

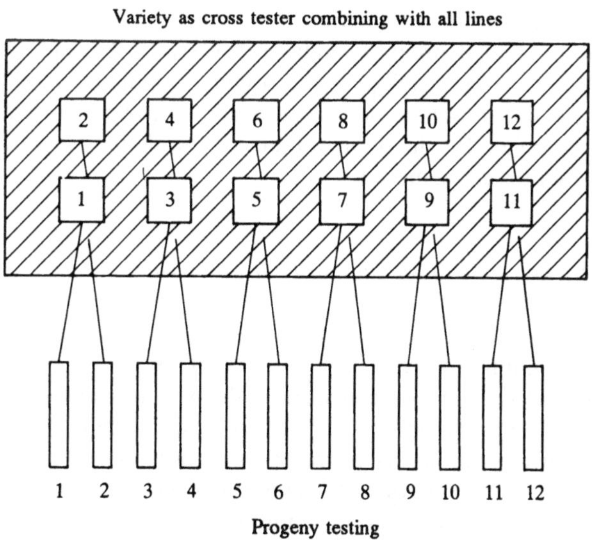

Fig. 20. Topcross

initially be produced. At the same time, original material is formed that is especially favorable for the development of I lines.

In *recurrent* selection, plants from a population are selfed and, after the yield of the selfed seeds, the progenies of the phenotypically best plants are cultivated in the 2nd year. The phenotypically best progenies are then crossed in as many combinations as possible and the seeds obtained hereby are cultivated in the 3rd year as a population. Within this already improved population, selection and selfing of plants take place again and, with this, the second cycle of the recurrent selection begins (Fig. 21).

In *reciprocal recurrent* selection, two populations (A + B) form the basis for the selection. Reciprocally, one population serves as tester for the examination of the selections deriving from the other population. From population A, a greater number of plants is selfed and at the same time crossed with several plants of B. The same is done with population B. In the 2nd year, the progenies of the testcrosses are subjected to a performance trial. The progenies of the cross of *one* female parent, which originated from one population, are combined with several male parents from another population. Based on the results received by the tester experiments, progenies are cultivated in the 3rd year from selfed seed produced by plants that displayed outstanding performance in the performance trial of the testcrosses. Furthermore, during this 3rd year, the best progenies of selfings are crossed in as many combinations as possible. With the seeds obtained in this way, improved populations are formed in the 4th year. Within these populations individual plants are again selfed and crossed and, herewith, the new cycle in recurrent reciprocal selection begins (Fig. 22).

This method leads to an accumulation of genes, not only for valuable specific characteristics but also for combining ability. By constantly crossing selfed progenies with one another that have good general combining ability, too strong a limitation of the genetic variability is avoided and, thereby, favorable original material for the production of inbred lines serving as parents for hybrid varieties is supplied. In order to obtain a heterosis effect as large as possible in the F_1 hybrid that is to be developed, the original populations for the reciprocal recurrent selection must be genetically different from one another. These methods can also be applied for the improvement of populations serving as original material for other breeding methods.

2.3.1.4 *Seed Production of Hybrid Varieties*

At the end of a hybrid breeding program, there are two lines with good specific combining ability, which must be crossed with one another for hybrid seed production. In most cases, one of the two lines serves as female line (seed parent), which must be pollinated by the second one. Fertilization of the female line with its own pollen is to be prevented, because this leads to inbred and poorly vigorous plants, and consequently counteracts the advantages of the hybrid variety. For this reason problems arise, initially of a technical nature, during production of

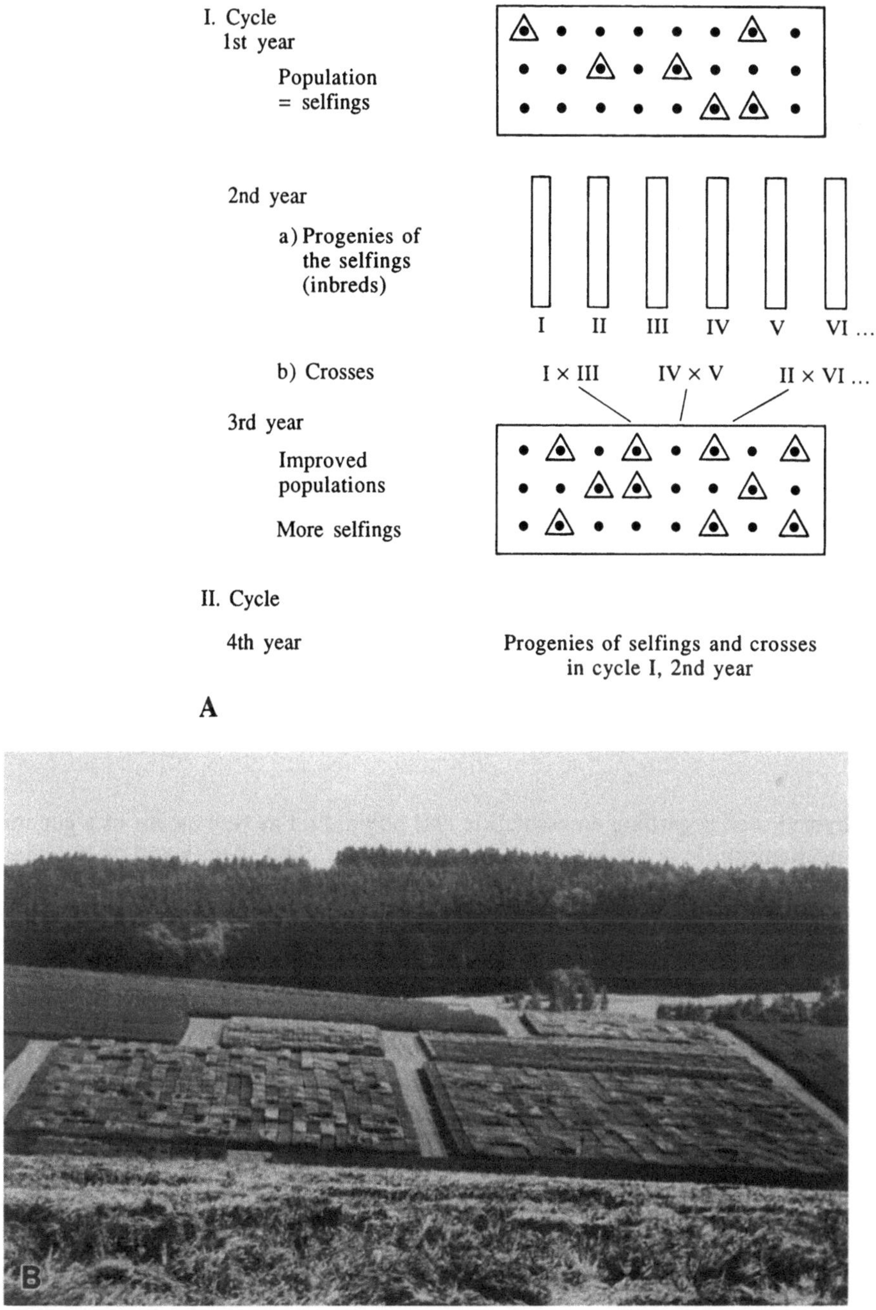

Fig. 21. *A* Diagram for recurrent selection; *B* depiction of practical application – experimental plots for testing resistances against fungi in wheat

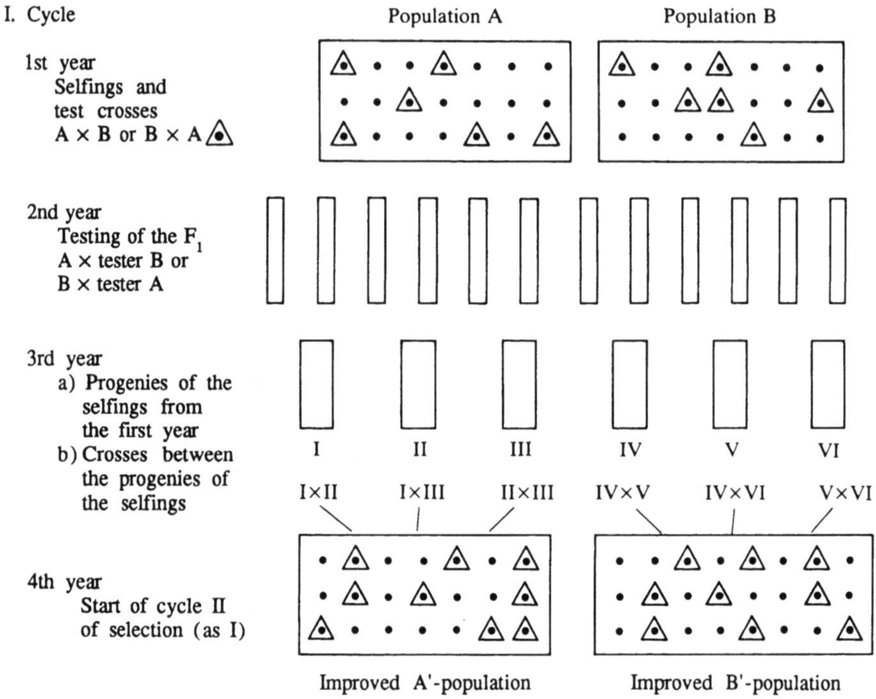

Fig. 22. Diagram for reciprocal recurrent selection

hybrid seed regarding emasculation and pollination as well as use of a genetic mechanism; these are largely determined by the plant species and its breeding system. Economic problems also occur, since the additional costs for seed production are only justified insofar as they can be compensated by a correspondingly higher profit in relation to conventional varieties. Thus, the method to be applied also depends on the price relation of hybrid seed to standard seed. The price of hybrid seed can amount to about 5 times more in spinach, 8 in onion, 35 in pickling cucumber, 40 in peppers, and even 120 in tomato compared to conventional varieties. Thus, in tomato, for instance, in contrast to spinach, much more laborious and costly methods of hybrid seed production can be employed.

Some of our cultivated plant species, such as maize, cucumber, and begonia, are monoecious. They form separate male and female flowers distributed on one plant. If the distribution of the female and male flowers on one individual as well as the chronological sequence of their maturity are favorable, mass emasculation and the production of hybrid seed cause no difficulties.

In maize, for instance, which is a classic example for monoecism in hybrid seed production, the male flowers are very favorably placed on a tassel at the top of the plants, so that they can easily be removed by hand (detasseling). For F_1 hybrid seed production the inbred lines A and B are cultivated next to each other in rows. Generally, there is one row of the male line to three to five rows of the female line (Fig. 23). Only the A plants are detasseled, so that they produce cross seed A × B for sale. Propagation of the A and B lines takes place in different isolated fields. If the I lines have only low intrinsic performance, the F_1 seed yield will be only mediocre. This disadvantage can be obviated, as has been done on a large scale in maize hybrid breeding, if from two single crosses, e.g., A × B and C × D, a double cross (A × B) × (C × D) is produced. The double cross seed can be obtained in the same way as the seed from single crosses by detasseling the female partner (Fig. 24). Due to the vigor of the single crosses brought about by heterosis, the seed production of a double cross is correspondingly high. The double hybrids can also be high yielding, if the right F_1 hybrids are combined. With respect to uniformity, however, the double hybrids are inferior to the single crosses. Since over the course of years the performance of the I lines could be considerably improved, due in part to the methods described in the preceding section, the production of three-way hybrids or even single hybrids is economically feasible. Three-way hybrids are cross seed resulting from a combination of an F_1 progeny with a suitable partner (e.g., I line) serving as pollinator.

From 20 I lines, for example, except for the reciprocal crosses, the following can be carried out:

190 single crosses [n(n-1)/2],
3420 three-way crosses [n(n-1)(n-2)/2], and
14 535 double crosses [n(n-1)(n-2)(n-3)/8].

With such a great number of possibilities, experimental tests of all combinations are not feasible. However, if the influence of epistasis on the heterosis effect can be ignored, it is possible to calculate the yield of a double or three-way cross, based on the mean yields of the single cross, which do not directly participate in the double or three-way cross. For example, from the four I lines (A, B, C, D), excluding reciprocal crosses, six single crosses can be made (A × B, A × C, A × D, B × C, B × D, C × D). The yield of the double cross (A × B) × (C × D) can be calculated by the mean yield of the four single crosses A × C, A × D, B × C, B × D. Anderson (1938) confirmed this in an experiment with maize, using five inbred lines resulting in 10 single crosses and 15 double crosses. The actual yields of the double crosses showed only slight differences from the calculated yields based on the single crosses. [An example drawn from this experiment: actual yield (bushels per acre) of the double hybrid (A × B) × (C × D) = 68.8; yields of the single hybrids, which are not involved in this double hybrid, were A × C = 62.6, A × D = 70.8, B × C = 65.6, B × D = 72.1; mean yield of the four single hybrids = 67.8.] Using quantitative genetic models, the validity of this method of precalculation can be explained.

In the dioecious plant species, e.g., spinach (*Spinacia oleracea* L.), asparagus (*Asparagus officinalis* L.), and hemp (*Cannabis sativa* L.), the male and female flowers grow on different individuals. In populations of such plant species, male and female plants occur at a ratio of 1:1. Roguing of male plants in a dioecious crop, and using the remaining female plants as seed parents after pollination with a corresponding male line, is theoretically possible. This is provided, however,

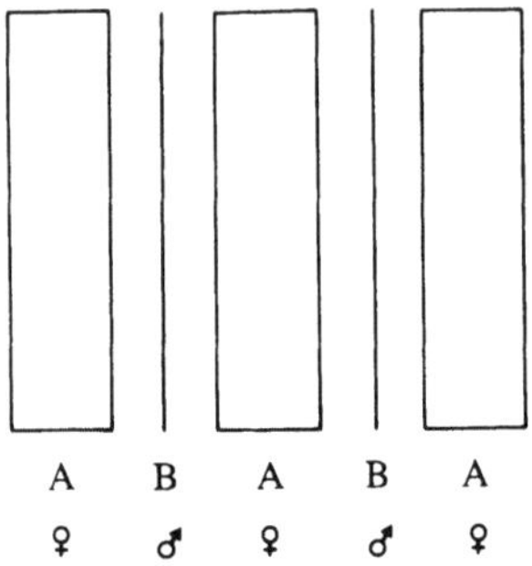

Fig. 23. Production of single cross hybrids in maize; hybrid seed is harvested from the detasseled female line A

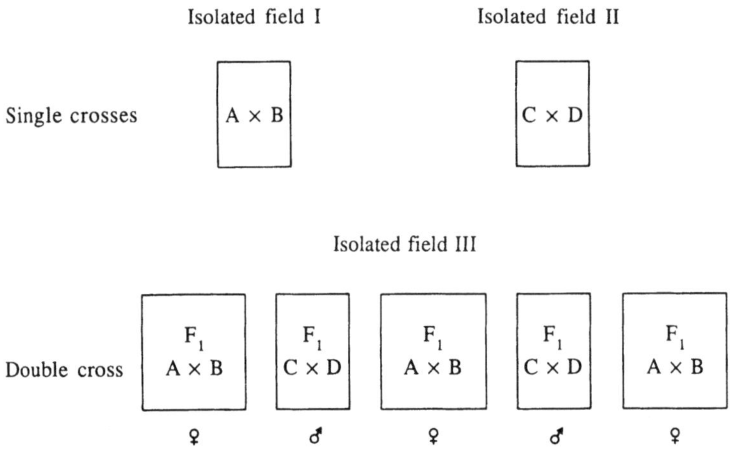

Fig. 24. Production of double cross hybrids in maize; hybrid seed is harvested from the detasseled female line A × B

that the undesired male plants are identified and removed before their pollen is dispersed. In spinach as well as in asparagus, this method is used for hybrid seed production (see Sect. 3.5).

Contrary to the monoecious and dioecious plant species, however, most of our cultivated plant species are hermaphroditic, where male and female sex organs are combined in one flower. Due to the biology of the flower, a mass emasculation of the seed parent by hand and subsequent hand-pollination is possible only in very few plant species (such as tomato, pepper, eggplant, petunia), and hybrid seed production based on this is very seldom economically feasible. In such plant species, there are various genetic mechanisms for the production of hybrid seed, which have recently been compiled and discussed by Wricke (1989). Next, these mechanisms will be described in connection with hybrid seed production.

One of these mechanisms is self-incompatibility. This signifies the inability of plants to set seed after selfing, in spite of functioning gametes. In nature, this mechanism prevents self-fertilization in higher plants and sustains cross-pollination, which is of special importance for the maintenance of genetic variability and heterozygosity in allogamous plant species. As already mentioned in Section 2.3.1.1, there are two kinds of self-incompatibility: "heteromorphic" and "homomorphic" self-incompatibility.

Within *Primula*, a genus with heteromorphic self-incompatibility, types with long styles, the filaments of which are shorter than the style (kk), and types with short styles having correspondingly longer filaments are distinguished (heterostyly). Short styles are monogenic dominant over long styles and phenotypes showing this character are heterozygous (Kk). Besides these morphological differences, a physiological mechanism such as sporophytic incompatibility, which will be discussed further below, prevents self-fertilization, but mutual fertilization of Kk with kk genotypes and vice versa is possible and results in a 1:1 segregation. Thus, the short-styled types can be rogued in one hybrid parent and the long-styled types in the other and, after mutual pollination, hybrid seed can be harvested from both parents. This method is used in *Primula vulgaris* and in *Primula elatior* for the production of hybrid seed. However, seed production takes place at a time when no insects are available, so that pollinations must be done manually in an isolated greenhouse.

In homomorphic self-incompatibility, there is no morphological difference between the flower organs of different plants, but a genetically conditioned incompatibility between style or stigma tissues, respectively, and pollen grain prevents fertilization. It is caused by physiological factors and may be compared to the effect of an antigen-antibody reaction. Two types are distinguished: "gametophytic" and "sporophytic" self-incompatibility. In general, both systems are genetically controlled by a series of multiple alleles of one locus, which are termed S_1, S_2, S_3, etc. The number of alleles can be high, for example, in *Brassica* species, it amounts to about 50. Gametophytic incompatibility systems with two loci (in some grass species, Lundquist 1956) and with four loci (*Beta vulgaris*, Larsen 1974) have also been described.

The gametophytic system of incompatibility is found in important cultivated plant species such as sugar beet, potato, tobacco, clover, and grasses. In the gametophytic system, the incompatibility of the pollen is determined after meiosis, so that only the genotype of the pollen is a deciding factor. The pollen tube growth of a pollen is prevented or slowed down in the style if the pollen has an S allele that is also in the diploid stigma or style tissue of the receiving plant. In the crossing of S_1S_2 with S_2S_3 only the S_3 pollen can fertilize, whereas S_2 pollen tube growth is therefore prevented. Consequently, only two combinations result from this cross: S_1S_3 and S_2S_3. Plants with identical alleles are mutually sterile (intrasterile), otherwise they are interfertile. If the parent plants differ in two S alleles, four intrasterile but interfertile groups are formed in the offspring, such as from the cross $S_1S_2 \times S_3S_4$ the genotypes S_1S_3, S_2S_3, S_1S_4, and S_2S_4.

The sporophytic system of incompatibility occurs in important cultivated plant species such as cabbage, broccoli, sunflower, and cocoa. In this system, incompatible pollen is already recognized at the stigma surface and its germination is prevented or, if it germinates, formation of the pollen tube is restricted. The basic difference of the sporophytic compared to the gametophytic system is that the compatible or incompatible reaction of the pollen is already determined prior to meiosis in the diploid tissue of the sporophyte of the male plant. The inheritance of two S alleles of a heterozygous plant may be codominant or they may interact with one another in such a way that they weaken each other, or one allele may also be dominant over the other. These relationships can be different in anther and stigma tissue. The dominance of one allele over the other may lead to important consequences for plant breeding. If, for example, in the anther tissue of an S_1S_2 plant S_1 is dominant over S_2, the S_2 pollen behaves as S_1 pollen, so that both will be incompatible with an S_1S_3 plant, but compatible with an S_2S_3 plant. If it is assumed that S_1 is dominant over S_2 and S_3, and S_2 over S_3 in the anther tissue, and that there are no dominance relationships between those three alleles in the style tissue, i.e., they are all equally effective. In contrast to the gametophytic system (see above), the cross $S_1S_2 \times S_2S_3$ will be sterile; this is because the pollen is determined by S_2 and is therefore incompatible with the stigma tissue. The reciprocal cross, however, is fertile, since the pollen, determined by S_1, is able to fertilize. The progeny will consist of S_1S_2, S_1S_3, S_2S_3, and S_2S_2 genotypes, thus also a homozygous type appears. Normally, such reciprocal differences and the formation of homozygous individuals resulting from such crosses are not possible in gametophytic self-incompatibility, unless selfing takes place due to pseudofertility.

Pseudo-compatibility occurs when the self-incompatibility mechanism is weakened or rendered ineffective, which may be caused by modifying genes, certain environmental, or also physiological factors such as flower age and plant age. It is found in the gametophytic as well as in the sporophytic system. The strongly modifying influence of environmental factors on self-incompatibility is of great importance for breeding. Thus, seed set can be obtained, for example, in *Brassica oleracea* by bud pollination. At this stage, as well as in aging flowers, the incompatibility mechanism is not yet or is no longer effective. To develop homozygous inbred lines, self-incompatibility can also be overcome by increasing or decreasing temperature, increasing atmospheric humidity or CO_2 concentration, cutting or damaging the stigmas and, finally, by grafting compatible stigmas onto incompatible plants.

In cole crops (*Brassica oleracea* L.), hybrid seed production makes use of the sporophytic incompatibility mechanism. Due to the great number of alleles, cabbage populations generally consist of plants that are heterozygous at the incompatibility locus. With bud pollinations, inbred lines may be developed from such a population. If I lines with the constitution S_1S_1 and S_2S_2, which have been tested for specific combining ability, are produced by this technique and cultivated next to one another, only seeds with the allele combination S_1S_2 will be formed; this is because the S_1 pollen cannot fertilize any egg cells of the S_1S_1 plants, nor can S_2 pollen fertilize any egg cells of the S_2S_2 plants. To prevent occurrence of pseudo-fertility and pollination of parent lines by selfing during hybrid and seed production, it is necessary to develop I lines with strongly dominant alleles.

If the seed production after a single cross is too low and therefore uneconomical, double crosses can also be carried out according to the following scheme:

$$S_1S_1 \times S_2S_2 \qquad S_3S_3 \times S_4S_4$$
$$S_1S_2 \quad \times \quad S_3S_4$$
$$S_1S_4,\ S_2S_4,\ S_1S_3,\ S_2S_3$$

Then, four I lines with good combining ability instead of two must be found and maintained. As the result of the pseudo-fertility, e.g., due to high temperatures or late flowers, pollinations within the parent lines leading to the development of selfed or sibbed seeds, respectively (which shall both be referred to as "sib"), must always be taken into account. Depending on the cole species, the plant material, environmental conditions, and modifying genes, their proportion may amount to as much as 40%. In addition to this, mutations leading to a self-fertility allele (S_F) occasionally occur, rendering the self-incompatibility mechanism ineffective. "Wrong" alleles, which enter the population uncontrolled, may considerably disturb the developed system, so that a strict supervision of the self-sterility in the developed material is indispensable. Seed companies are forced to ascertain the sib proportion in each hybrid seed lot, which must not be more than 5% in EC countries and in the USA. In general, this is done by cultivation of a representative seed sample under field conditions until the sib portion can be positively recognized. Experimental results of isozyme analysis for manifestation of the proportion of sibs in cole hybrids have been reported in several publications (Wills and Wieseman 1980; Arus et al. 1982). Application of recessively inherited waxlessness, which has recently been found in kohlrabi, has also been suggested for determination of sibs in hybrid seed lots as a marker.

As already mentioned above, self-incompatibility in grass species, and consequently in rye, is gametophytically controlled by two loci, S and Z. If the alleles of the two incompatibility loci present in the pollen also occur in the plant that is to be pollinated, no fertilization takes place. On the other hand, fertilization is not prevented if only one of the two loci has the same allele. England (1974) suggested that use be made of this gametophytic incompatibility mechanism in hybrid seed production of forage grasses, especially between closely related species such as *Lolium perenne* and *Lolium multiflorum*. However, this requires production of progenies from selfed plants. In rye, this is possible by application of high temperatures (30 °C) during flowering. If a plant that is heterozygous for both loci, e.g., $S_1S_2Z_3Z_4$, is selfed and its progeny is left isolated for open-pollination for two generations, a population develops that is in equilibrium and in which one-half of the plants cannot be fertilized by the pollen of this population (50% of the plants: $S_1S_1Z_3Z_4$, $S_2S_2Z_3Z_4$, $S_1S_2Z_3Z_3$, $S_1S_2Z_4Z_4$, and 50%: $S_1S_2Z_3Z_4$, which cannot be fertilized by the pollen of the population). If two such populations, being in equilibrium and containing different S and Z alleles, are subject to open pollination, theoretically a hybridization amounting to 83.3% instead of the usual

50% occurs. It could be demonstrated in experiments with rye that this outcrossing rate is almost reached. This method of hybrid seed production by use of gametophytic incompatibility is still in the experimental stage. Due to the considerably lower labor necessary in the development of parent lines as well as hybrid seed production, however, this method could gain practical importance if certain technical questions, e.g., concerning multiplication of the parent lines, which must be free of contamination, are solved (Wricke 1985).

Male sterility is the genetic mechanism that is most often used for hybrid seed production. A plant is called male sterile if it cannot produce functioning pollen. The ovaries of such plants generally function normally, so that they set enough seed after pollination with pollen of fertile plants.

In earlier publications three groups of male sterility are distinguished: "genic," "cytoplasmic," and "genic-cytoplasmic" male sterility. The majority of scientists dealing with male sterility in recent years do not tend to distinguish between the two latter forms but rather merge them into the term "cms," which is the abbreviation for "cytoplasmic male sterility." Cms describes the type of male sterility that is controlled by a cytoplasmic factor and mostly only one nuclear gene, Ms/ms (restorer gene). The male-sterile plants have the cms cytoplasm (S) and are recessive at the restorer locus: (S)msms. In the (S) cytoplasm the fertility is restored by the dominant restorer gene. All plants with normal cytoplasm (N) are male fertile, independent of the constitution at the restorer locus. The genetic constitution of the male-fertile plants can be (S)MsMs, (S)Msms, (N)MsMs, (N)Msms, or (N)msms. It is generally believed that the cytoplasmic factor responsible for male sterility or fertility is not transferred by the pollen. Thus, it is possible to produce progeny consisting of male-sterile plants only by crossing (S)msms × (N)msms, because the offspring do not inherit factor (N) of the male parent, but only the sterility factor (S) of the female parent, and female as well as male plants produce gametes with the recessive ms allele only.

Such progeny, consisting of male-sterile plants only, can be pollinated by a male line with good specific combining ability for hybrid seed production. This form of hybrid seed production is used in the most important cultivated plant species, such as maize, sorghum, pearl millet, sunflower, sugar beet, onion, and others. Since the genotypes containing the normal cytoplasm, recessive at the restorer locus, serve for multiplication or maintenance of the male-sterile line, they are termed maintainer genotypes.

In some plant species, e.g., onion or chives, cytoplasmic male-sterile plants may occur in varieties that are already being cultivated. Their frequency varies according to plant species and variety. There are also plant species, such as spinach (*Spinacia oleracea*), in which no cytoplasmic male-sterile plants have yet been found (Kobabe 1972). If the breeder has found male-sterile plants in his crops, he must try to find maintainer genotypes, (N)msms, via a number of test crosses with male-fertile plants. In plant species that easily set seed after selfing, such as onion, pair crossings between male-sterile and -fertile plants can be carried out. If the progeny resulting from such a pair crossing consists of male-sterile plants only, the corresponding pollinator is a maintainer and its seed obtained by selfing can be used for multiplication of the maintainer. In plant species, however, that are self-sterile, as in sugar beet, a self-fertility allele S_F may be transferred into the material from which maintainers are being selected. Otherwise, it is recommended that one male-sterile plant be pollinated with two fertile plants and that further selection work be concentrated on the progeny of the fertile plants, which already lead to a satisfactory number of male-sterile plants after the first crossing. The frequency of maintainer genotypes varies, as already mentioned, according to plant species and population. In onion, for example, such genotypes occur with an average frequency of 5%, whereas maize populations consist predominantly of such genotypes.

If there are no spontaneously arising cms plants in a plant species, it can be possible to obtain such plants by mutagenic treatments of the seeds. Also, after intergeneric and interspecific crosses, cms plants can be developed. New techniques of cell and protoplast fusion and cell regeneration are additional tools for creating cms plants.

In hybrid varieties of plant species that are used vegetatively, such as sugar beet, onion, forage grasses and others, the genetic constitution of the male hybrid parent is of no importance. If, on the other hand, an (N)msms type is used as pollinator in plant species from which generative parts, such as seeds, are being used (maize, sunflower), a male-sterile F_1 progeny is obtained which, of course, cannot produce any seeds. In such species, the male lines must restore the male fertility of the hybrid. This will succeed if the male parent is of the type (S)MsMs or (N)MsMs, whereby an (S)Msms F_1 progeny can be produced, being male fertile, due the Ms allele present in every plant. A line containing the dominant restorer gene and restoring fertility in the F_1 hybrid is called a restorer line. Finding a pollinator with good restorer abilities may be more or less difficult, depending on the plant. In species in which restorer genes do not often occur, the already identified restorer genes may be added to inbreds by a series of backcrosses, using the inbred as the recurrent parent.

Consequently, three components are necessary for the production of an F_1 hybrid when using cms (Fig. 25): (A) a cms line, (B) a maintainer line for maintenance and multiplication of the cms line, and (C) a pollinator line with good specific combining ability, which must also be able to restore fertility in plant species whose generative parts are being used. It is possible, too, to produce three-way as well as double hybrids by application of cms (Fig. 26a,b). In order to develop a three-way hybrid, first a male-sterile F_1 hybrid must be produced. This hybrid must be crossed with another pollinator line, which may also be an inbreeding line. For the production of a double hybrid, one of the two F_1 hybrids must be male sterile, so that it can be pollinated by the second F_1 hybrid. If the second F_1 hybrid was also produced with the aid of cms and consists of plants with a heterozygous restorer locus, fertility is only restored in one-half of the plants of the double hybrid, whereas the other half is male-sterile. This is acceptable in allogamous plant species producing sufficient quantities of pollen, e.g., maize, because the pollen quantity produced by one-half of the plants is enough for the fertilization of all the plants. In maize, however, there is also the possibility of producing one of the F_1 hybrids by detasseling the corresponding female line, using for this female and male lines with a dominant restorer locus.

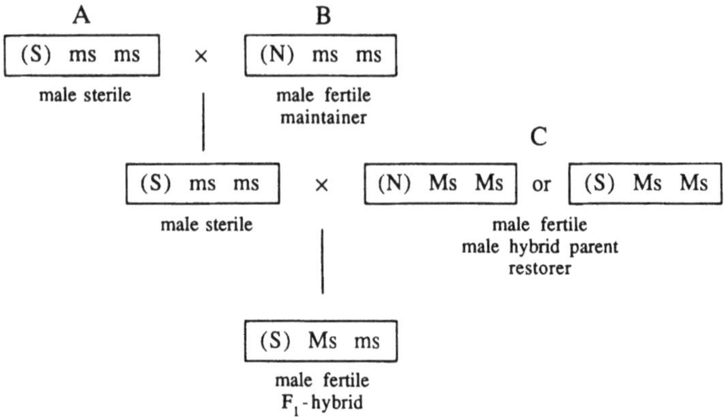

Fig. 25. Production of a single cross hybrid using CMS

As a consequence, an F_1 hybrid is obtained that is homozygous dominant at the restorer locus (Fig. 27). If this pollinates another single cross that was produced using cms, a double hybrid is obtained, consisting of restored plants only. It is possible to produce the second single cross with a homozygous restorer gene also by use of genic male sterility.

In some plant species, more than one restorer gene is necessary for the restoration of fertility in the male-sterile cytoplasm. In cms S cytoplasm in sugar beets and in cms T cytoplasm in maize, for example, there are two complementary dominant genes, each responsible for the restoration. Within a plant species, different types of cms cytoplasm can occur, reacting to different restorer genes. Since cytoplasmic uniformity may imply vulnerability, occurrence of different cms cytoplasms in a plant species and their usage in hybrid seed production is advantageous. In 1970, for example, maize cultivation in the USA broke down because of a southern corn leaf blight epidemic. At that time, almost all of the cultivated maize hybrids in the USA were based on the use of the cms T cytoplasm. It was found that the epidemic had been caused by race T of the fungus *Helminthosporium maydis* (*Drechslera maydis*), which was only virulent in plants with the T cytoplasm.

In some plant species, the expression of cms is influenced also by environmental factors. Some cms plants in chives (*Allium schoenoprasum*) become male fertile under constant temperatures of 24 °C. This fertility is reversible. Under normal temperatures (20 °C day/14 °C night), these plants flower male sterile again. The temperature reaction is controlled by a dominant nuclear gene (Tatlioglu 1985). The cms plants, the temperature gene of which is recessive, are male-sterile under all temperature conditions. It is very probable that such specific genes causing the temperature sensitivity of cms also exist in other plant species. They could permit multiplication of a cms line without a maintainer, and thus could be used in hybrid breeding. The precondition for their usage is that hybrid seed production can be carried out under controlled temperatures, e.g., in a greenhouse.

Some cms plants in chives are sensitive to tetracycline; they become male fertile after they have been sprayed with a solution containing tetracycline. Tetracycline sensitivity, which is also reversible, is controlled by a recessive nuclear gene (Tatlioglu and Wricke 1988). As in some other species, there are indications that in chives, too, a specific mitochondrial polypeptide (or polypeptides) are correlated with cms. It is possible that synthesis or the effect of the cms-specific polypeptide in the corresponding sensitive cms plants is disturbed, or that its effect is restricted by constantly high temperatures or tetracycline.

Genic male sterility, which can be found in many cultivated plant species, such as wheat, maize, sorghum, sugar beet, pearl millet, tomato, and others, is usually inherited by a recessive nuclear gene designated as st. Its usage in plant breeding is possible on a limited scale only. If such stst genotypes, for example, occur in a crop, natural pollination with StSt genotypes will lead to heterozygous, fertile Stst plants segregating in 25% of male-sterile plants after selfing or pair crossing. Two-thirds of the remaining population consist of Stst plants and one-third of StSt genotypes. If plants are subject to open-pollination in consecutive generations and seed is only harvested from male-sterile plants, a population is formed in the F_4 generation, consisting of 50% male-sterile stst genotypes and 50% heterozygous Stst genotypes (Table 12), because in the F_2 population only the homozygous dominant and heterozygous plants produce pollen that is able to fertilize. The ratio of the male-fertile genotypes, 1/3 StSt to 2/3 Stst, leads to a frequency of 0.66 for the St allele and of 0.33 for the st allele in the F_2 population which, after open-pollination within the progeny of the male-sterile plants, results in an F_3 population consisting of 33.3% stst plants and 66.6% Stst plants. One-half of the fertile pollen of this population now contains the St allele, the other half

Table 12. Using genic male sterility in plant breeding

Generation	Genotypes			Allele frequency in male gametes
	♀		♂	
Parents	stst	×	StSt	1.0 St : 0.0 st
F_1	Stst (self-pollinated)			0.5 St : 0.5 st
F_2	25% stst	×	25% StSt + 50% Stst	0.66 St : 0.33 st
F_3	33.3% stst	×	66.6% Stst	0.50 St : 0.50 st
F_4	50% stst	×	50% Stst	0.50 St : 0.50 st
F_5	50% stst	×	50% Stst	0.50 St : 0.50 st

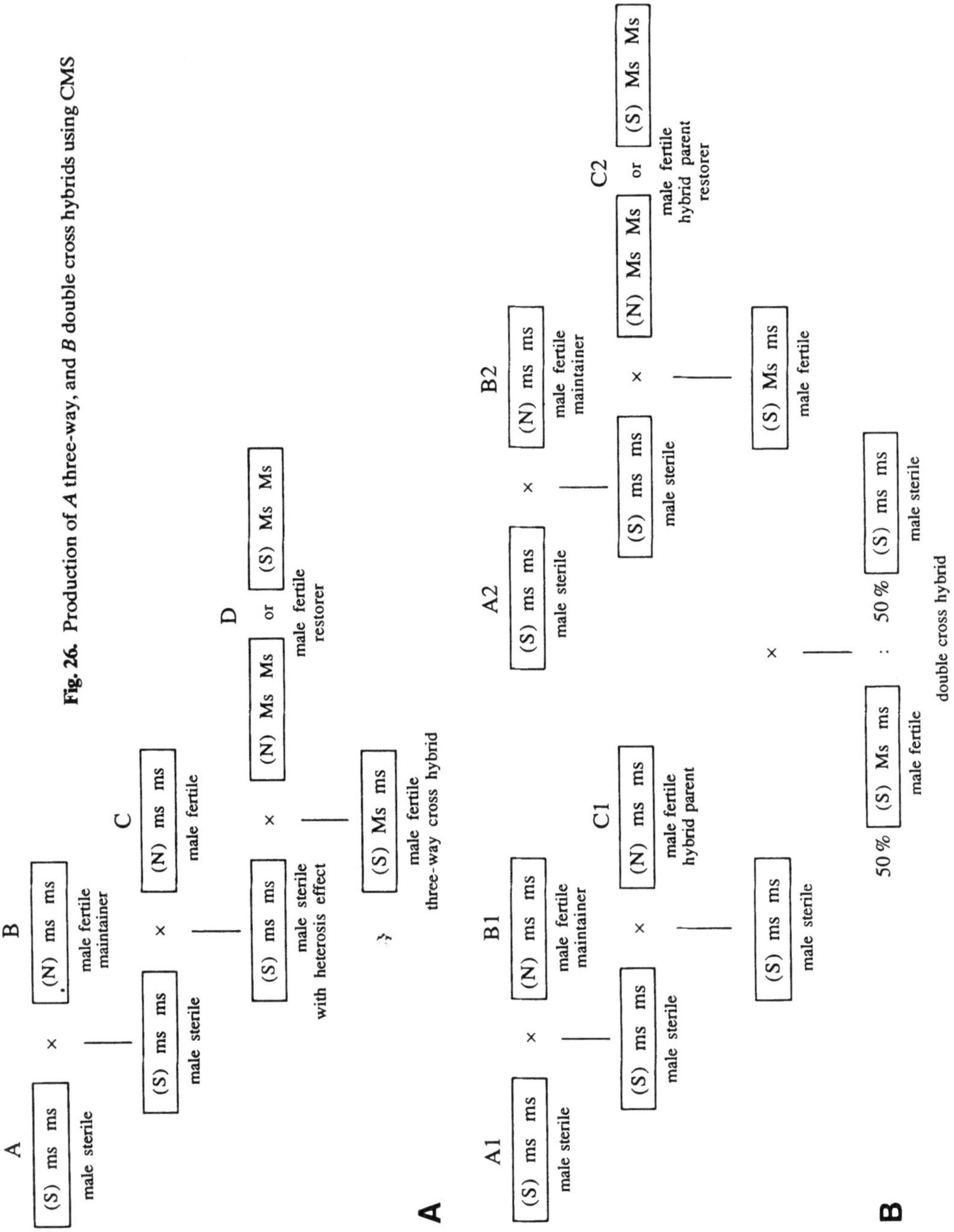

Fig. 26. Production of *A* three-way, and *B* double cross hybrids using CMS

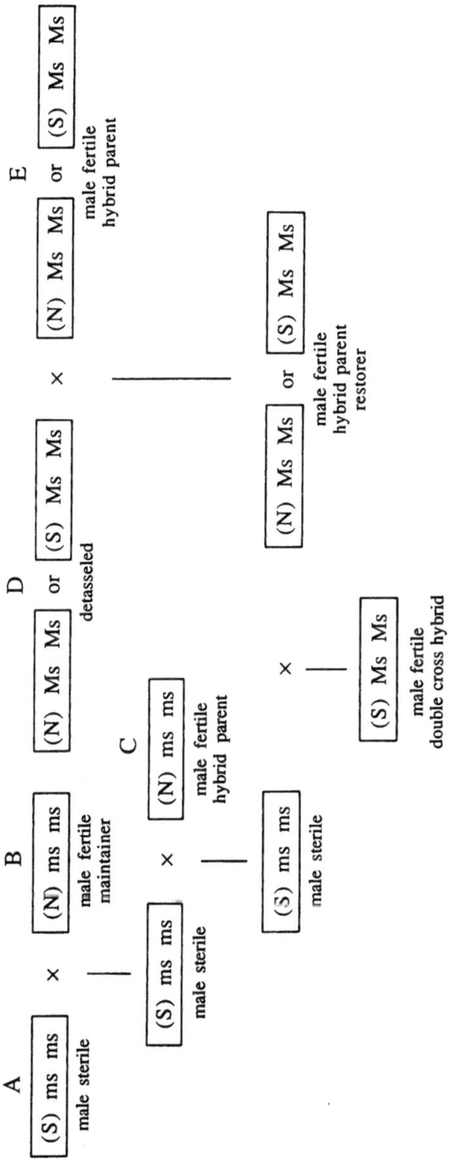

Fig. 27. Production of a double cross hybrid using CMS and manual emasculation (detaselling)

the st allele, so that the offspring of the male-sterile plants of this population lead to an F_4 progeny already consisting of 50% stst plants and 50% Stst plants.

The development of such a population can be advanced if the male-sterile plant originally found can be successfully propagated vegetatively. Thus, it can be backcrossed with the F_1 hybrid Stst, which has been produced in the meantime. This cross stst × Stst results in 50% male-fertile Stst plants and 50% male-sterile stst plants. In the next generation, the progeny of male-sterile plants from such a population segregates into male-sterile and -fertile plants again at a ratio of 1:1 after open-pollination. In this way, the genic male sterility can be multiplied at a constant frequency of 50% male-sterile plants. If the male-fertile plants can be successfully rogued from the field before pollination, the male-sterile plants can be used as female lines for a hybrid variety. Since, contrary to cytoplasmic male sterility, development of lines consisting of male-sterile plants only is not possible with genic male sterility; this form of male sterility can only be used in plant species in which no cytoplasmic male sterility useful for practical breeding could be found or generated. Due to the additional labor required for roguing of the fertile plants, application in breeding is only lucrative when relatively small amounts of hybrid seed are necessary and when a high price can be obtained (e.g., *Ageratum, Carthamus, Chrysanthemum, Tagetes, Zinnia,* etc.). The increase of the portion of male-sterile plants or the reduction of hand labor in the roguing of male-fertile plants is possible if the St allele is linked to a marker gene, the phenotype of which is already expressed in the seedling stage or essentially earlier than flowering. In tomato, for example, two genes [Prx-2 (Peroxidase-2) and aa (anthocyanin absent)], which are linked to the ms-10 gene, are suggested for this purpose (Tanksley and Zamir 1988).

In addition to self-incompatibility, male sterility, monoecism, and dioecism, apomixis or agamospermy can also be used for hybrid seed production. In agamospermous species, fully functioning seed is formed on the female plant without fertilization of the egg cell having necessarily taken place. Consequently, plants resulting from such seed are absolutely identical with the female parent, so that in this manner, a specific genotype can be identically propagated by seed. If a plant is characterized by an especially high performance after a cross, this can thus be fixed by agamospermy. However, a cross between apomictic species is only possible if the agamospermy is not mandatory. There are quite a number, particularly of subtropic grass species, in which this kind of fixing of the heterosis effect has been practically tested. In the People's Republic of China, a new apomictic rice variety, "84–15," with high yield, good tolerance to stress, and sequential use of heterosis has recently been developed.

If none of the methods for production of hybrid seed so far described can be applied due to their expense, or if the flower biology of the corresponding cultivated species does not permit production of F_1 seed in quantities usual in commerce, the production of so-called chance hybrids, e.g., in forage grass breeding, can be considered. For this, seed of the hybrid components is mixed at a ratio of 1:1 and cultivated for seed production. According to the laws of population genetics, the seed yield of this mixed crop should consist of 50% pure F_1 hybrid seed, whereas the other half is a combination of the seed of both components. Under certain circumstances, it can be assumed that after sowing such a chance-hybrid variety, the more vigorous F_1 plants will suppress all the others, so that after some time all nonhybrids will have disappeared (Kobabe 1983).

2.3.2 Hybrid Breeding of Self-Fertilizing Species

The great successes of hybrid breeding in cross-pollinated species such as maize, onion, millet, ornamental plants, and others led to the employment of this breeding method also in self-fertilizing species. It's problems and technical procedures, however, differ in many respects from those of cross-pollinated species. An increase in performance is also quite often observed in F_1 combinations in self-fertilizing species. By selections in subsequent generations, genotypes can be bred in self-fertilizing species, which transgress their parents in performance with regard to quantitative as well as qualitative characters, whereas in cross-fertilized species, such a true breeding of heterosis observed in F_1 progenies can generally not be achieved.

The question arises as to whether the total heterosis in the F_1 can be bred to constancy in a self-fertilizing species, or if part of it is genetically controlled in such a way that true breeding is not possible, e.g., in heterosis, which is conditioned by superdominance of one or several genes.

The same also applies to the effect of multiple alleles, if a combined effect and thus an increased expression of a character or production of a specific substance takes place due to the specific effect of two different alleles of one gene locus. An increased F_1 performance can be bred to constancy only if the genes combined within the F_1 of a heterotic hybrid exhibit dominant or intermediate inheritance, if no very tight linkage prevents their combination, and if no inbreeding depression occurs in the subsequent generations.

It may be asssumed with great probability that part of the F_1 heterosis in self-fertilizing species is so genetically conditioned as to make true breeding impossible, and thus development of hybrid varieties seems useful. Even without this assumption, the development of hybrid varieties seems to be justified as opposed to conventional varieties in certain cases. Besides the natural variety protection of a hybrid, occasionally the breeding aim of an F_1 hybrid is also realized more quickly and effectively than by selection and testing of the subsequent generations for several years. Breeding of the F_1 hybrid for its part presupposes economical seed production for cultivation. In tomato breeding, it is easier to realize polygenic characters, such as early maturity and high yield or high early yield, within one *hybrid variety* than by breeding of a conventional variety with constancy in these features. For the desired combination of a quantitative character, such as fruit size, with a dominant resistance gene to a parasitic fungus, development of a hybrid variety is also to be preferred.

In hybrid breeding of self-fertilizing species, there is no development of inbreeding lines and thus no difficulty of maintenance. Testing for general combining ability can also be omitted if the breeder restricts himself to diallel crosses of a limited number of existing varieties or genotypes from his genetic stocks. The technique of seed production depends on whether mass emasculation and manual pollination on a large scale are practicable and above all economical (as is, e.g., the case in tomato), or whether only natural wind pollination is possible in the application of a cytoplasmic, pollen-sterile female line, e.g., in wheat. Today, inducing pollen sterility with chemicals (gametocides) is still under discussion. An even more recent approach to male sterility uses gene transfer techniques: a gene for ribonuclease production is coupled to an anther-specific promotor, thus inactivating microspores in the anther by a strong RNase overproduction.

In tomato, emasculations can quickly and easily be carried out by hand. After hand-pollination of one single flower, a fruit containing about 100 seeds is produced. The use of male-sterile types does not offer very great advantages since, so far, only purely genically conditioned pollen sterility is known and thus – as has already been described above – the continuation of such lines by simultaneous breeding of 50% male-fertile plants (Msms) involves additional costs. If it were possible to find a linkage between ms and a dominant marker gene, which already manifests itself at an early stage of development, the use of male-sterile female plants would be more attractive. If emasculated or male-sterile flowers are left to natural pollination, only a small number of fruits is produced per plant and seed set is strongly reduced. The technique of hand-pollination can be improved by suitable collection and preservation of the pollen.

The development of hybrid varieties in wheat is based on the discovery of male-sterile types by Kihara (1967). By introducing a genome of tetraploid and hexaploid cultivated wheat, *Triticum aestivum* ssp. *vulgare* or *T. durum*, into the plasm of *Triticum timopheevi, Aegilops caudata*, or *Aegilops ovata* with the help of several backcrosses, pollen-sterile, cultivated wheats with a purely female inheritance are produced; after pollination with a normal variety, a purely pollen-sterile progeny is formed again. Certain genetic and breeding problems arise in the attempt to use such male-sterile types for the development of hybrid varieties. They are related to the development of suitable female and male lines and the production of hybrid seed.

The female line must not only have a male sterility that is as stable as possible, i.e., not easily modified to fertility by environmental influences, but at the same time it must have a high intrinsic performance and, above all, good specific combining ability with the male line. Since the interaction of alien plasm/genome often has a negative effect on the expression of characters compared to the effect of the genome in the original plasm of its own species, the genomes of numerous varieties must be introduced into the alien plasm, until usable male-sterile genotypes with sufficient intrinsic performance and combining ability are found. The male variety must have restorer genes that restore pollen fertility again in the F_1 hybrid. Restorer genes only rarely occur worldwide. If they are found, they must be introduced into the genome of a suitable pollinator line by several backcrosses. A further difficulty in the development of suitable male lines is that the effect of the restorer genes with regard to the restoration of fertility is quite variable and is also determined by the genotype of the female line, and that environmental conditions may also modify their effect. The amount of seed set on the male-sterile female line is largely dependent on the amount of pollen distributed from the male line by air and, furthermore, on the flight distance of the pollen and the duration of its ability to fertilize. The degree to which wheat varieties flower openly, thus enabling their pollen to be distributed by air, differs depending on genotype as well as environment. In arid areas, an open-pollination is very much favored by protrusion of the anthers. Compared to rye, the amount of pollen is smaller in wheat and its range is shorter; however, it is sufficient for pollination if enough anthers protrude. Its considerably shorter viability could be a greater disadvantage. Thus, the male line of a hybrid variety must also be selected with these characters in mind. It may be assumed that there is genetic variation in the characters important for cross-fertilization, so that improvements are possible by means of selection.

For propagation of the pollen-sterile female line, Ramage and Wiebe (see "Induced Mutations in Plants", Vienna 1969) developed a new method, which is based on the experimentally induced structural change of an additional chromosome. This method is discussed in Section 3.1.2, as more detailed knowledge on the induction of mutations is necessary. It should already be pointed out here that the special significance of this method is that it is based on the use of a purely genic pollen sterility controlled by Mendelian genes. This kind of pollen sterility can be found more often in cultivated species than the cytoplasmic type. In this new method, the difficulties connected with the identification of the restorer genes for seed production disappear.

2.4 Synthetic Varieties

Due to their high yield potential and uniformity of their yield products, hybrid varieties presently have an increasing distribution and are therefore also of growing economic significance for world nutrition. The high costs required, not just for their development but also for hybrid seed production, are only feasible if a correspondingly higher price than for conventional varieties can be achieved on the seed market, and if there is a higher turnover for large cultivation areas. Further preconditions for successful hybrid breeding, including seed production, are the presence of a staff of skilled technicians and such technical facilities as well-managed, agricultural propagation farms. If these prerequisites are not fulfilled, as is the case in the emergent plant breeding of some developing countries, breeding of synthetic varieties should take the place of hybrid breeding. In their potential performance, if not always in their uniformity, synthetic varieties can be equivalent to hybrid varieties. However, their second generation can be cultivated by untrained persons without significant loss of yield.

As is demonstrated below, synthetic varieties are rather an exceptional case of breeding of cross-pollinated species, because as in allogamous, open-pollinated varieties, advantage is taken of heterosis due to general combining ability. Also, the seed for market for synthetics is produced in the same manner as in the varieties of cross-pollinated crops; there is no strict regulation of pollination as in the production of hybrid seed. On the other hand, the methods for selection of the parent genotypes to be used are similar and, finally, the motivation of the breeder to develop synthetic varieties is often based on the fact that, if hybrid breeding proves to be too difficult and laborious, at least the production of synthetic varieties has been started. Thus, it seems justified to relate synthetic varieties to hybrid varieties and that their description is included in the chapter dealing with hybrid breeding.

The development of synthetic varieties begins with a selection of genotypes that are not only tested for their performance per se but also for high combining ability. Original material for selection of a synthetic variety are I lines, clones, and selected families. The genotypes that have proven to be of high combining ability are used as components for the synthetic variety. The generation in which the components are combined for mutual pollination is designated as Syn 0 and the generations following panmixis, as Syn 1, Syn 2, Syn 3, etc. The method of their development by selection of I line and testing for combining ability is in principle the same as in hybrid varieties. Their main difference from hybrid varieties is their propagation by panmixis. While the hybrids from single crosses reach the highest degree of heterozygosity in the F_1, which decreases again in the course of subsequent generations, the heterozygosity and performance potential achieved in Syn 1 can largely survive in the following generations, too, if several components are used for synthesis.

The testing for combining ability is the decisive criterion for a synthetic variety by which it can be distinguished from a conventional variety of a cross-pollinating species, which originates in continuous selection of individuals and subsequent progeny tests. If several lines of largely the same phenotype are combined in breeding a self-pollinating species such as wheat in order to render the variety more adaptable and thus more reliable in yield, a mechanical mixture is obtained, which should by no means be termed a synthetic variety. Nevertheless, variety mixtures may have slightly higher yields than separately cultivated varieties due to compensation phenomena.

The greater variability caused by crossing several components with high combining ability makes the synthetic varieties more flexible and thus also more adaptable compared to conventional varieties. Similar to hybrids, the yield generally also decreases after the Syn 2 in synthetic varieties, until an equilibrium is reached which, in partially self-fertile species, depends on selfing rate and inbreeding minimum, but mainly also on the number of components used in the Syn 0. The performance of Syn 1 can be estimated according to the formula:

$$\hat{F}_2 = \bar{F}_1 - \frac{\bar{F}_1 - \bar{P}}{n}$$

(Allard 1960). The yield to be expected usually increases with the number of components until an optimum is reached. The question as to the most favorable number of genotypes the Syn 0 should be composed of cannot be clearly answered, because the evidence from research and practice is too divergent.

The yield can be increased by:

1. an increase of the mean performance of the F_1 combinations,
2. an increase of the mean performance of the parents (I lines, clones),
3. an optimum combination of the components (out of 16 selected components only one 16-synthetic can be developed, but 12 870 8-component synthetics,some of which may widely surpass the performance of the 16-synthetic).

Therefore, it is obvious that I_0 plants should already be tested for their combining ability and that I_1 plants or lines should be used as components, the inbreeding depression of which is not yet as strong as in the I_5.

In order to maintain performance in subsequent generations, mass selections have proven to be sufficient in maize. The performance of the synthesis can be improved by one further breeding cycle. It consists of selection of genotypes from the synthetic variety, their testing by a diallel, and the combination of genotypes with the highest combining ability for a new synthesis (= recurrent selection).

Synthetic varieties were first successfully developed using the above-described methods in maize. In fodder plants (grass and clover species), improved varieties were successfully developed to an increasing extent using a slightly modified procedure, the "polycross."

In Section 2.3.1.3, this method has already been indicated as a tool for the testing of genotypes for their general combining ability. Since the polycross has not only proven to be a very effective breeding method in fodder plants but also in other cultivated plants, especially those that can also easily be vegetatively propagated, such as chives and others, this method of Frandsen will be described in more detail here (Fig. 28). From a crop of 5000 to 10 000 individual plants originating from material modified by inbreeding and other methods, about 500 are selected and multiplied in clones of 20 cuttings each. Based on repeated testing and observation of these 500 clones, 60–100 clones are selected for the establishment of the polycross. For this, each clone is divided into 100 cuttings and planted in 20 different places with five plants each within the polycross, in order to realize the most important principle, namely, that each clone is pollinated by a pollen mixture containing pollen from all clones, if possible. Therefore, the replications of the clones are distributed at random. The seeds of all replications of a clone are harvested together and this mixture is subject to a progeny test, with replications at two or three different locations. Based on these tests, 5–20 clones displaying the best combining ability are used for development of a synthetic variety. After another transplantation and vegetative propagation, these selected clones are mixed with each other and planted on a larger field for production of the elite seed.

For the production of a polycross, a simultaneous flowering of all components is necessary, because only in this way is a good mixture of the pollen and, consequently, a uniform and varied genetic mixture and combination rendered possible. Also, as many replications of the genotypes as possible contribute to this; the plant number may be reduced to one individual each per replication.

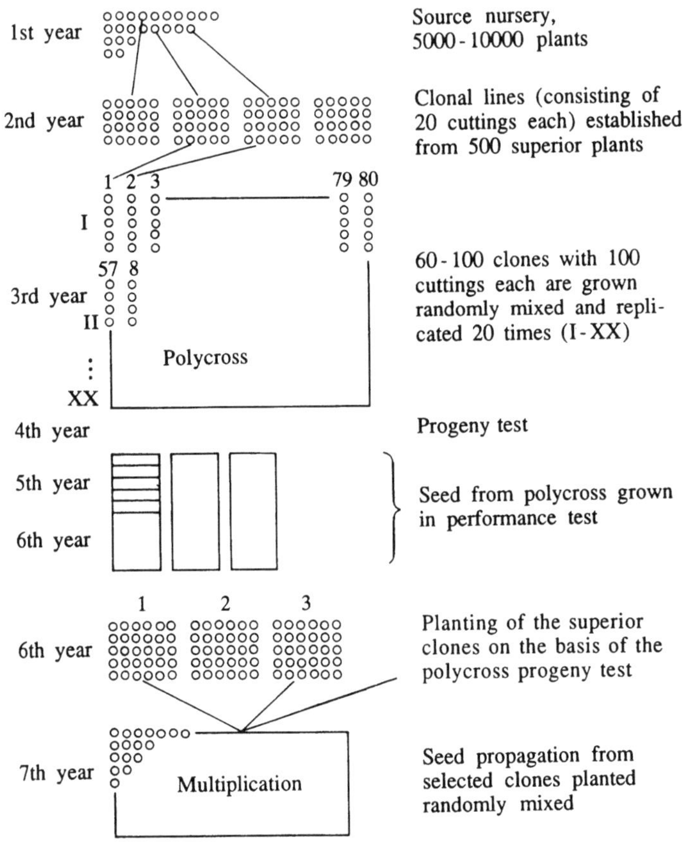

Fig. 28. Diagram of the polycross method (Frandsen and Frandsen 1948)

If the classical method of individual selection with progeny tests is used in cross- fertilized species, a more or less strong inbreeding depression takes place in continuous selection of individual plants within the same family or closely related families. This results in extensive uniformity of the selected characters which may, however, at the same time be asociated with a more or less strong yield depression. Since in a genetic respect the phenomena of inbreeding and heterosis are very closely related to each other in cross-fertilized species, an attempt should be made to compensate for the inbreeding effect by combining several homogeneous families with subsequent cross between crops. Whether this leads to as large an increase of performance as in synthetic varieties, in which the components are selected for a cross between crops based on their combining ability, is questionable.

2.5 Cited and Recommended Literature for Chapter 2

Allard RW (1966) Principles of plant breeding. Wiley & Sons, New York

Arus P, Tanksley SD, Orton TJ, Jones RA (1982) Electrophoretic variation as a tool for determining seeds purity and breeding hybrid varieties of *Brassica oleracea*. Euphytica 31:417–428

Becker HC (1982) Züchtung synthetischer Sorten. II. Leistungsvorhersage und Selektion der Eltern. AG Pflanzenzücht Heft 1:23–40

Becker WA (1975) Manual of quantitative genetics. Students Books Corporation, Pullman, Washington

Blakeslee AF, Belling J (1924) Chromosomal mutations in the Jimson weed. *Datura stramonium*. J Hered 15:195–206

Bolz G (1968) Monohybride Vererbung der plasmatisch-genisch bedingten Pollensterilität bei *Beta vulgaris* L. Z Pflanzenzücht 60:219–234

Briggs FN, Knowles PF (1967) Introduction to plant breeding. Reinhold, New York

Chase SS (1963) Analytical breeding in *Solanum tuberosum* L. – a scheme utilizing parthenotes and other diploid stocks. Can J Genet Cytol 5:359–363

De Buyser J, Henry Y (1986) Florin: a doubled haploid wheat variety developed by anther culture method. In: Somers DA et al. (eds) IAPTC Abstracts. Univ Minnesota, Minneapolis, p 416

De Nettancourt D (1977) Incompatibility in angiosperms. Monogr Theor Appl Genet 3. Springer, Berlin Heidelberg New York

Dickson MH, Wallace DH (1986) Cabbage breeding. In: Basset MJ (ed) Breeding vegetable crops. Avi, Westport, pp 395–432

Duvick DN (1965) Cytoplasmic pollen sterility in corn. Adv Genet 13:1–56

East EM (1908) Inbreeding in corn. Rept Connecticut Agric Exp Sta, pp 419–428

Edwardson JR (1970) Cytoplasmic male sterility. Bot Rev 36:341–420

England FJW (1974) The use of incompatibility for the production of F_1 hybrids in forage grasses. Heredity 32:183–188

Falconer DS (1981) Introduction to quantitative genetics. Longman, London New York

Foster CA (1971) Interpopulational and intervarietal hybrids in *Lolium perenne*; heterosis under simulated sward conditions. J Agr Sci 76:401–410

Foster CA (1973) Prospects for the development of F_1 hybrid varieties of perennial ryegrass (*Lolium perenne*) and other herbage grasses. J Agr Sci 81:33–47

Frandsen HN, Frandsen KJ (1948) Polycross-Methoden. Nord Jordbruks Forsk 7–8:239–261

Frankel R, Galun E (1977) Pollination mechanisms, reproduction and plant breeding. Monogr Theor Appl Genet 2. Springer Berlin Heidelberg New York

Fyfe IL, Gilbert L (1963) Partial diallel crosses. Biometrics 19:278–286

Guha S, Maheshwari SC (1964) In vitro production of embryos from anthers of *Datura*. Nature 204:497

Harlan JR, Martini LM (1938) The effect of natural selection in a mixture of barley varieties. J Agr Res 57:189–199

Hermsen JHT, Verdenius J (1973) Selection from *Solanum tuberosum* group *phureja* of genotypes combining high-frequency haploid induction with homozygosity for embryo-spot. Euphytica 22:244–259

Ho KM, Jones GE (1980) Mingo barley. Can J Plant Sci 60:279–280

Hoffmann W, Mudra A, Plarre W (1971) Lehrbuch der Züchtung landwirtschaftlicher Kulturpflanzen 2 Bde, Bd 2 (1985), 2nd edn, Parey, Berlin

Horn W (1974) F_1 hybrids in floriculture. Proc XIXth Int Hort Congr, Warszawa IV:267–277

Hurd RG, Biscoe PV, Dennis C (eds) (1980) Opportunities for increasing crop yields. Pitman, Boston

Jian-Sun C (1988) A great breakthrough in rice breeding in China. People's Daily, Beijing, Quesseas Edition, Oct 5th (translation by Tang SX)

Johannsen W (1909) Elemente der exakten Erblichkeitslehre. Fischer, Jena

Jones DF (1917) Dominance of linked factors as a means of accounting for heterosis. Genetics 2:466–479

Kappert H (1953) Die vererbungswissenschaftlichen Grundlagen der Züchtung. Parey, Berlin

Kappert H, Rudorf W (1958–1962) Handbuch der Pflanzenzüchtung. Bd I-VI, Parey, Berlin

Kasha KJ, Kao KN (1970) High frequency haploid production in barley (*Hordeum vulgare* L) Nature 225:874–875

Kaul MLH (1988) Male sterility in higher plants. Monogr Theor Appl Genet 10. Springer, Berlin Heidelberg New York
Kermicle JL (1969) Androgenesis conditioned by a mutation in maize. Science 166:1422–1422
Kihara H (1967) Cytoplasmic male sterility in relation to hybrid wheat breeding. Züchter 37:86–91
Kobabe G (1972) Die Vererbung der männlichen Sterilität beim Spinat (*Spinacia oleracea* L.) und Möglichkeiten der Nutzung dieser Eigenschaft in der Hybridzüchtung. Z Pflanzenzücht 67:233–242
Kobabe G (1983) Heterosis and hybrid seed production in fodder grass. In: Frankel R (coord ed) Heterosis, Monogr Theor Appl Genet 6. Springer, Berlin Heidelberg New York, pp 124–137
Lapusher D, Frankel R (1967) Practical aspects and the use of male sterility in the production of hybrid tomato seed. Euphytica 16:300–310
Larsen K (1974) Four loci governing the self-incompatibility in *Beta vulgaris* L. Incomp Newsl Assoc Euratom-Ital, Wageningen 4:20–22
Londsdale DM (1987) Cytoplasmic male sterility: a molecular perspective. Plant Physiol Biochem 25:263–271
Lundquist A (1956) Self-incompatibility in rye. I. Genetic control in the diploid. Hereditas 42:293–348
Melchers G, Labib G (1970) Die Bedeutung haploider höherer Pflanzen für Pflanzenphysiologie und Pflanzenzüchtung. Ber Dtsch Bot Ges 83:129–150
Nitsch CN, Sheridan W, Neuffer G. Anderson S, Godard M (1982). The production of haploids from maize and millet. In: Earle E, Demarly Y (eds) Variability in plants regenerated from tissue culture. Praeger Publ, New York, pp 69–91
Owen FV (1945) Cytoplasmically inherited male sterility in sugar beets. J Agric Res 71:423–440
Pearson OH (1983) Heterosis in vegetable crops. In: Frankel R (coord ed) Monogr Theor Appl Genet 6. Springer Berlin Heidelberg New York, pp 138–188
Pike LM (1986) Onion breeding. In: Basset MJ (ed) Breeding vegetable crops. Avi, Westport, pp 357–394
Poehlman JM (1987) Breeding field crops, 3rd edn. Avi, Publ Van Nostrand Reinhold, New York
Poole CF (1956) Improvement to tomato ascorbic acid content. Proc Am Soc Hort Sci 68:443–451
Ramage RT, Wiebe GA (1969) Use of chromosome aberrations in producing female parents of hybrids. In: IAEA (ed) Induced mutations in plants. Wien, pp 655–660
Redei GP (1982) Genetics. MacMillan, New York, Collier MacMillan London
Reimann-Philipp R (1983) Heterosis in ornamentals. In: Frankel R (coord ed) Heterosis. Monogr Theor Appl Genet 6. Springer, Berlin Heidelberg New York, pp 234–259
Rieger R, Michaelis A, Green MM (1976) Glossary of genetics and cytogenetics, 4th edn. Springer, Berlin Heidelberg New York
Ross H (1986) Potato breeding -- problems and perspectives. Adv Pl Breed Suppl 13. Parey, Berlin
Schmalz H (1989) Pflanzenzüchtung. VEB Deutscher Landwirtschaftsverlag, Berlin
Schnell FW (1977) Züchtungssysteme zur Leistungssteigerung bei Nutzpflanzen. Hohenheimer Arbeiten, Allgem Reihe Heft 89, Ulmer, Stuttgart
Schuster W, Geidel H (eds) (1978) Der landwintschaftlich-technische Assistent. Ulmer, Stuttgart
Shull GH (1908) The composition of a field of maize. Rep Am Breeder's Assoc 4:296–301
Simmonds NW (1979) Principles of crop improvements. Longman, London
Tanksley SD, Zamir D (1988) Double tagging of a male-sterile gene in tomato using a morphological and enzymatic marker gene. Hort Sci 23:387–388
Tatlioglu T (1985) Influence of temperature on the expression of cytoplasmic male sterility in chives (*Allium schoenoprasum* L.). Z Pflanzenzücht 94:156–161
Tatlioglu T (1989) Inheritance of waxlesness in kohlrabi (*Brassica oleracea* L. var. *gongloides*) and it's utilization in hybrid seed production. Plant Breed 102:215–221
Tatlioglu T, Wricke G (1988) Genetic control of tetracycline sensitivity of cytoplasmic male sterility (cms) in chives (*Allium schoenoprasum* L.). Plant Breed 100:34–40
Tatum LA (1971) The southern corn leaf blight epidemic. Science 171:1113–1116
Ullstrup AJ (1972) The impact of the southern corn leaf blight epidemics of 1970–71. Ann Rev Phytopathol 10:37–50
Vose PB, Blixt SG (eds) (1984) Crop breeding -- a contemporary basis. Pergamon Press, Oxford
Wenzel G, Debnath SC, Schuchmann R, Foroughi-Wehr B (1987) Combined application of classical and unconventional techniques in breeding for disease resistant potatoes. In: Jellis GJ et al. (eds) Prod New Potato Var -- Techn Adv, Univ Press, Cambridge, pp 277–288

Wenzel G, Schieder O, Przewozny T, Sopory SK, Melchers G (1979) Comparison of single cell culture derived *Solanum tuberosum* L. plants and a model for their application in breeding programs. Theor Appl Genet 55:49–55
Wills AB, Wieseman EM (1980) Acid phosphatase isoenzymes of *Brassica oleracea* seedlings and their application to sib testing in F_1 hybrids. Ann Appl Biol 94:137–147
Wilson P, Driscoll CJ (1983) Hybrid wheat. In: Frankel R (coord ed) Heterosis. Monogr Theor Appl Genet 6. Springer, Berlin Heidelberg New York, pp 94–123
Wricke G (1985) Hybrid breeding in rye using self-incompatibility. Eucarpia Cereal Section Rye, Svalöv, Proc Part I:267–277
Wricke G (1989) Genetic mechanisms for hybrid seed production. Proc XII. Eucarpia Congr Göttingen, Parey, Berlin, pp 369–378
Wricke G, Weber WE (1986) Quantitative genetics and selection in plant breeding. De Gruyter, Berlin
Yordanov M (1983) Heterosis in tomato. In: Frankel R (coord ed) Heterosis. Monogr Theor Appl Genet 6. Springer, Berlin Heidelberg New York, pp 189–219

3 Special Breeding and Selection Techniques

Special procedures have been developed, and in some cases are still in the process of being developed, to render selection more effective or to solve special problems. Genetic expansion of the original material for selection also serves to strengthen breeding methods. Thus, the production and use of mutants, including those of polyploid types, are discussed in Sections 3.1.1 to 3.1.3. Breeding research in these areas, which began in the mid-1920s, has contributed to substantial progress in practical breeding. More recently it has been discovered that in cell and tissue cultures, spontaneous variations are formed during vegetative propagation of a genotype, the phenotypic changes of which are not always as drastic as is the case in the majority of induced mutants. Furthermore, it is possible in cell and tissue cultures mutagenically to treat and select a large number of haploid and diploid cells in the smallest space (Sect. 3.1.4).

Genetic changes in periclinal chimeras, i.e., types in which the cell layers (L_1, L_2, and L_3 layer) are of different genetic constitution, are also closely linked with the problem of mutation induction. Therefore, rearrangement of the cell layers, which occurs spontaneously or may also be induced by mutagenic treatment, can change the expression of characters (Sect. 3.2).

In Sect. 3.3 the development and practical significance of apomixis is discussed in more detail. In this type of propagation, seed set occurs without the fusion of two gametes and often also without reduction division. The resulting possibilities of permanently maintaining a heterozygous genotype also in seed multiplication have already been pointed out in Section 2.3.1.1, on the features of inbreeding and heterosis. In this chapter, the first principles of reproductive biology necessary for understanding the processes involved in the production and use of haploids by parthogenesis and androgenesis are given. These are also discussed in more detail in Section 3.4. The fundamentals of the unconventional methods explained here are important again in conjunction with the conventional procedures discussed in Sections 3.5, 3.6.3, and 3.7. In a broader sense, the description of species and genera hybrids (Sect. 3.6.1) pertains to methods of combination breeding (Sect. 2.2); it is discussed here because their preparation in breeding requires knowledge of allopolyploidy and, as an alternative method, protoplast fusion, which is possibly much more effective due to the fusion of two genomes and two plasmons of different genetic constitution.

3.1 Production and Use of Mutants

Mutations are changes in the genotype, hence changes of a hereditary nature, which are expressed by a different reaction norm of the mutated plants compared to the nonmutated original plants. They include formation of morphological characters as well as the physiological behavior as a function of environmental factors. The process of hereditary change is termed mutation and the individuals affected by the change are called mutants. Besides recombinations, mutations are the cause of genetic variation, from which every breeding activity proceeds. By means of a mutation, the multiplicity of types of a population is not only increased by *one* genotype, but also by a large number of further new genotypes due to the possibility of subsequent recombinations of the mutant with the existing genotypes.

To use mutations in breeding successfully, it is necessary for the breeder to familiarize himself with the different forms of mutations, their development, manifestation, and genetic behavior and, above all, to acquire the necessary knowledge to increase the frequency of their spontaneous occurrence by experimental intervention.

The different forms of mutations can be classified with regard to the structures within the cell:

(a) *Genome mutations*: they comprise the mutations within the nucleus and can be subdivided as follows.

1. Point mutations.
2. Chromosome mutations, i.e., changes in chromosome structure.
3. Ploidy mutations, i.e., changes in chromosome number.

(b) *Plastidome mutations*: the plastids also contain heritable elements that control the function of the plastids and sometimes other traits and characteristics of an individual, too.

(c) *Plasmon mutations*: hereditary changes of the cytoplasm.

Since the cytoplasm and the plastids are predominantly transmitted by the egg cell, the plastidome and plasmon mutations show a purely maternal inheritance. The inheritance controlled by plastidome and plasmon is termed female, plasmic, or extra-chromosomal inheritance. The manner in which the genes of the plasm and the nucleus work is only to be understood by their mutual interaction, as has been demonstrated, e.g., for pollen sterility determined by plasm.

3.1.1 Point Mutations

In classical genetics, genes have been defined as units of function for the formation of specific characters, as units of recombination, and as units of mutation, so that a gene mutation can be demonstrated in a Mendelian experiment; after a cross between the mutant and the original type, a monohybrid segregation takes place in the F_2. By linkage analyses the location of the mutation on the chromosome can be determined. Although in the light of more recent investigations, this definition can no longer be upheld in its entirety, it is still useful for an initial orientation regarding the manifold forms of mutations and their cytogenetic characterization. Here, it shall only be said that genes do not mutate as a unit. Such changes affect the nucleotides and consequently it is appropriate to use the term "point mutation" instead of "gene mutation."

Analogous to the presence of the chromosomes in pairs in a diploid organism, each gene locus is also represented by two alleles, which are designated by the symbols AA or aa, if identical, and by Aa, if differing from one another. In general, mutation only takes place in one allele; for the most part, it passes from a dominant allele to a recessive one: an AA genotype is changed to an Aa genotype and an Aa genotype to an aa genotype. If the mutation takes place in the generative tissue, a gametes are also formed in the orignal AA genotype, resulting in the generation of some Aa progenies. If A is completely dominant over a, the process of mutation after selfing will only be observed in the following generation, in which 25% of the aa genotypes segregate. If a mutation occurs within the somatic tissue, it can only be transferred to the progeny by sexual multiplication, if it has developed in the L_2 layer, from which the generative tissue is formed (Sect. 3.2). Somatic mutations in the L_1 can manifest themselves, e.g., in the flower, in the form of color-altered sectors. They can only be multiplied by vegetative propagation. By using cell culture techniques, e.g., protoplasts, it is possible to regenerate functional plants from just one layer which will contain three identical layers. If several characters have clearly changed at the same time, the mutation is called pleiotropic.

Spontaneous mutations occur in varying degrees in individual cultivated plants. Again, each cultivated species displays a certain variation in the mutation rate, which is influenced by the respective genotype and certain environmental factors, and which is specific for certain gene loci. Thus, especially in maize, specific genes, so-called mutator genes or controlling elements, could be identified that influence the degree of mutability of certain genes. Heterozygosity may also lead to the mutation of an allele and consequently to homozygosity of the corresponding gene locus.

The frequency of spontaneous mutations is too low to be able to be the basis for systematic breeding, even taking into account the internal and external factors promoting them. This does not change the fact that spontaneous mutations may be of great significance, whether found by chance, e.g., the dwarf mutant of castor beans in Brazil (from which highly efficient varieties for mechanical cultivation were developed), or by systematic search for certain mutants (as in the case of alkaloid-free forms and nondehiscing pods in lupine species).

It was not until about 60 years ago that the natural mutation rate was significantly increased by X-rays. Ionizing rays, such as X- and γ-rays and neutrons, from the group of corpuscular rays, have proven to be especially effective as mutagenic agents. Recently, an increasing number of chemical compounds with a mutagenic effect have become known, although only a few of them are appropriate for practical use in mutagenesis for breeding purposes. These are: ethyl methane sulfonate (EMS), ethylimine (EI), nitroethylurethane (NEU).

These are chemical compounds with active alkyl groups, which probably cause mutations by alterations in the molecules of the nucleotides, the carriers of genetic information.

Also tissue culture, particularly long-term tissue culture, induces mutations; due to its specific origin, this variation is called somaclonal – coming from calli; protoclonal – coming from protoplasts; or gametoclonal – coming from microspores (Sect. 3.1.4).

The technique of mutagenesis is first of all determined by the breeding system of the species in question. In species with generative propagation, preferably the seeds are subject to mutagenic treatment. First, it is important to determine the radiation dosage or concentration of the chemical compound, as well as the duration of the treatment that promises the greatest yield in usable mutations. The mutation rate increases with increasing dose; however, simultaneously the nongenetic primary damage increases as well, resulting in a decreasing ability to germinate, increasing inhibition of growth, and other morphological and physiological disorders during plant growth. For this reason, some authors recommend a weaker radiation dose, by which not more than 20-50% of the plants die off. With increasing dosage, the proportion of chromosome mutations with conspicuous morphological and physiological changes usually increases, too. They are generally of no significance for breeding purposes due to their simultaneously diminished vitality.

If a point mutation occurs in a barley seed, it can be expected that the plant deriving from this seed will be heterozygous at the mutated gene locus and that its progeny will segregate 25% mutants, insofar as the mutation is recessive. Analogous to the symbols used in Mendelian analyses, the first generation developing after a mutagenic treatment is called M_1 and the subsequent generation with the segregating recessive mutants is called M_2. If in the M_2 the individual ears deriving from the progeny of an M_1 barley plant are cultivated separately, segregation of a recessive mutation cannot – as in a Mendelian experiment with monohybrid inheritance – be observed in all ear progenies of the M_2, but only in one, seldom in several. Furthermore, the segregating progenies frequently have a deficit of recessives. These phenomena can be explained by the complex histogenetic structure of the embryo and its further differentiation into organs and tissues after irradiation of dormant seeds.

Within the embryo of the variety, the primordia of one main shoot and four lateral shoots are already preformed. They consist of a few (two to four) cells, from which the three tissue layers L_1-L_3 are differentiated. Only if the generative tissue layer (the L_2) is formed from a mutated initial cell do gametes develop containing the mutated gene locus; from their combinations the recessive mutants are obtained in the M_2. It is extremely improbable that in one initial cell of the five primordia the same mutation occurs and forms the L_2 simultaneously. The ears of the same plant may consequently have a different genetic constitution, which again explains the different behavior of their progenies. Megaspore and microspore mother cells of the L_2 of a heterozygous mutant Aa only form A and a gametes at a ratio of 1:1, if the L_2 is uniformly composed of Aa cells.

The above-mentioned mutagenic agents not only result in the induction of point mutations, but also in chromosome breakages, which may lead to the loss of fragments of individual chromosomes. Gametes with deficiencies are often lethal or diminished in their vitality, so that in competition they are inferior to normal gametes as regards the formation of zygotes. This elimination of gametes occurring in the haploid phase is called "haplontic selection." It is the reason that gametes are formed from a heterozygous plant in a numerical ratio deviating from that to be expected. Even gametes with an otherwise vital point mutation can be eliminated if they contain deficiencies. The fusion of gametes with deficiencies more often results in the formation of lethal zygotes, which are expressed in sterility of the corresponding flowers. The selection of gametes is followed by an elimination of zygotes, which is called "diplontic selection."

Since the majority of point mutations is recessive, the search for deviating phenotypes does not begin until the M_2 generation. The determination as to whether a detected variant is a mutant or a nonheritable modification can only take place in the M_3 after testing their progeny. In very rare cases only, e.g., after drastic changes of the phenotype, is it possible to prove the presence of a point mutation in the M_2 by identifiable Mendelian segregation.

The identification of point mutations, which are of special importance to the breeder, is very difficult. In a self-fertilizing species, the phenotypic expression of a quantitative character changed by a mutation generally lies within the natural mutation range of the normal original type, so that the segregating mutants cannot be recognized as such in the M_2 on inspection. Therefore, a large number of phenotypically normal plants must be selected from the M_2 and the individual plants must be cultivated as M_3 lines. Only if stands of the M_3 lines are compared to those of the original type can point mutations be verified, because crops of the M_3 plants can already be distinguished upon examination in characters such as early flowering or altered shoot length. The significance of the differences in quantitative features can be ascertained by applying biometrical methods.

The induction of mutations and the identification of mutants are followed by a test phase of their economic breeding value. To determine the breeding value, comparative tests must be carried out for several years under different environmental conditions. Thus, it is largely a matter of ascertaining, by means of a versatile experimental design, the interactions between genotype and the various environmental factors that have been changed by mutation.

Often the induced mutations display a pleiotropic effect, due to which several features are changed, only *one* of which, however, is desired by the breeder. This pleiotropic effect of a mutation may sometimes be interrupted by crossing the mutation into another variety, the genotype of which is largely different from the original type treated. The altered effect of a mutation on the development of characters in another genetic background can be caused by interaction with other genes. Also, the cross of two mutants can occasionally break undesired pleiotropic effects and valuable transgressions in other features can be achieved.

The procedures for induction of point mutations described so far refer to self-fertilizing cultivated plants, such as barley, oats, wheat, rice, peas, and others. In self-fertilizing species, it is not difficult to produce largely homozygous lines or varieties as original material for homozygous lines or varieties as original material for mutagenic treatment. The use of mutants in cross-pollinated species is made especially difficult if, after selfing, only insufficient seed set can be achieved due to self-sterility, or if the inbreeding damage of the progenies of selfings is too great. To avoid these difficulties, it is advisable that repeated mutational treatments be carried out.

The question is now posed as to whether or not induction of mutations is superior to combination breeding. An argument for its superiority is the shorter breeding process for a variety to be developed, and the possibility of improving in *one* character, a breeding variety with good performance.

The advantages in the use of mutations only partially apply, however. Also in conventional combination breeding, the breeding process can be considerably accelerated by application of the backcross method, if several generations are cultivated in 1 year under greenhouse conditions. If the desired character is present in a heritable form, normally the classical procedure is more successful, since in the backcross method the repeated recombinations will result in lines not only containing the new trait but also probably superior in a number of other characters compared to the starting material.

This final evaluation of the use of point mutations in breeding compared to other breeding procedures was mainly undertaken with regard to breeding sexually propagated cultivated plants. In vegetatively propagated, cultivated plants, the circumstances relating to induction and identification of mutations and their use in breeding are different from those in generatively propagated species, so that this group of cultivated plants requires special consideration.

Spontaneous mutations in somatic tissues, the bud mutations (sports), have resulted in valuable new varieties, especially in fruit species and diverse ornamental plants. An increase in the spontaneous mutation rate by mutagenic agents would increase the chance of the occurrence of desired mutations and accelerate the breeding process compared to the conventional method of developing novelties, namely, by cultivation of seedling progenies deriving from crosses or open-pollinated female plants. In species with long generation intervals, such as kernelled and stone fruit, tulips, and others, this would be a particular advantage. Another advantage of the induction of mutations is seen in the fact that the general character of the original variety is maintained in the sports, and that somatic mutations cannot be eliminated during meiosis by haplontic selection, as is the case especially in chromosomal structure changes and in aneuploids.

For induction of somatic mutations, only varieties of economic importance and with a high degree of heterozygosity should be used. Since mutations mainly change dominant alleles into recessive ones, a recessive mutation in the somatic tissue can only be phenotypically perceived if the original form is already heterozygous in the corresponding gene, i.e., if Aa mutates to aa. If A is dominant over a, a mutation from AA to Aa cannot be realized in somatic tissue. The mutagenic treatment takes place on those organs used in practice for propagation, such as scions, tubers, bulbs, rhizomes, and others. It is initiated in the resting stage of the buds or shortly prior to sprouting, at a time when the primordia consist of only a few, undifferentiated cells.

Mutations are only induced in one or in a few cells of a tissue, they are bound to one of the three layers forming the vegetative cone, and from which the individual organs are differentiated. From the mutated cell within a tissue layer first a mericlinal chimera is formed by anticline division and later on in the process a periclinal chimera, as can clearly be seen in Figs. 35 and 36 and in Section 3.2; that is, development proceeds from a labile state, in which the mutated tissue only appears as a sector, to the stable state of a periclinal chimera, in which a uniformly mutated tissue layer is present. Thus, periclinal chimeras are characterized by layers of different genetic constitution; they remain constant in vegetative propagation, in contrast to the mericlinal chimeras.

This histogenetic structure, combined with the competition between mutated and nonmutated cells in the primordia and already differentiated tissue layers, hinders the occurrence of somatic mutations. Small sectors with changed characters developing from an irradiated bud cannot easily be recognized, and the distinction between a mutation and primary damage due to irradiation is often very difficult.

The induction of somatic mutations in the breeding of vegetatively propagated ornamental plants has been successful. Reports on the accumulated induction of mutations in certain varieties of chrysanthemums, roses, and others should be taken with a degree of skepticism, however. In Section 3.2, it is pointed out that rearrangement of layers can be induced by X-rays. The resulting manifestation of new or changed characters is caused by the activation of latent or cryptomeric genes. Mutations can only be recognized if the gene in question is located within a layer, from which also the character controlled by this gene is formed. Thus, for example, mutation of an allele changing the color of an apple skin is only effective if this mutation occurs in the L_1, or if it reaches the L_1 by rearrangement of the layers. The new phenotypes, deriving from rearrangement of the layers in a periclinal chimera, can therefore not be regarded as the result of induced mutations, even if the layer rearrangement was due to mutagenic agents.

3.1.2 Structural Changes in Chromosomes

The genetic information originating in the genome is not only altered if single genes are changed at the molecular level, but also by structural changes of the chromosomes, which are the carriers of the genes.

The loss of fragments, termed "deficiencies" or "deletions," occurs after chromosome breakage, when the point of breakage fails to grow back together and the smaller end piece with a centromere is lost in the process of cell division (terminal deficiency; Fig. 31a). In the case of two simultaneous breakges, a middle piece of the chromosome may drop out and, after coalescence of the end pieces, a chromosome with a so-called intercalary deficiency is formed (Fig. 31b);

unequal crossing-over can also result in intercalary deficiencies. As can be seen from Fig. 29, the fragment that is lost in one chromosome is transferred to the other homologous chromosome, so that this produces a duplication in a specific segment. Deficiencies and duplications are initially heterozygotes, in the progeny of which homozygotes can also occur in connection with the structural change of a chromosome pair. Only if the deficiencies include smaller fragments of a chromosome are they viable as homozygotes. At times, they cannot be distinguished from point mutations in their phenotypic expression and in their genetic behavior. Duplications, on the other hand, are homozygously viable and also display changed character expressions.

After a double breakage, the released chromosome fragment can return to its earlier position. If the ends of the breakages join crosswise, however, an "inversion" is formed (Fig. 30). Pairing of an inverted chromosome with its homologous noninverted chromosome can result in complicated configurations (Fig. 31) that permit crossing-over within the inverted part of the chromosomes, but cause the dying off of the exchange gametes. In this respect, an inversion indirectly prevents the exchange of alleles. If the genes within the chromosome fragment relating to the inversion are heterozygous, this heterozygosity can remain permanently, provided that the inversion heterozygotes have selective advantage over the two homozygous forms, as has been demonstrated many

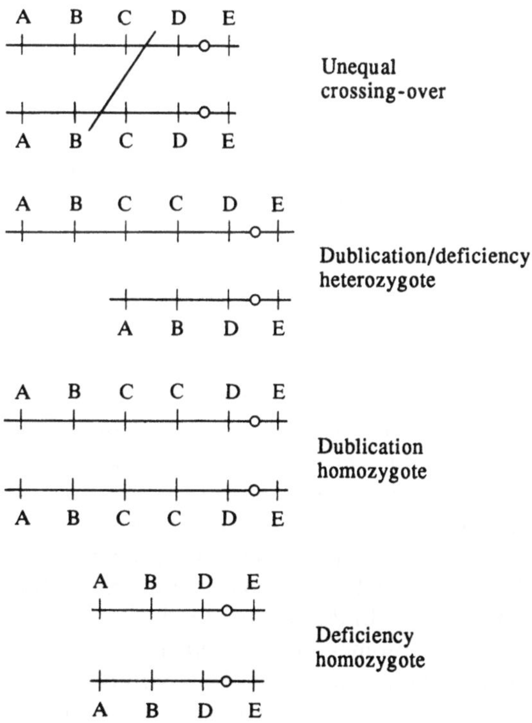

Fig. 29. Development of duplications and intercalary deficiencies by unequal crossing- over

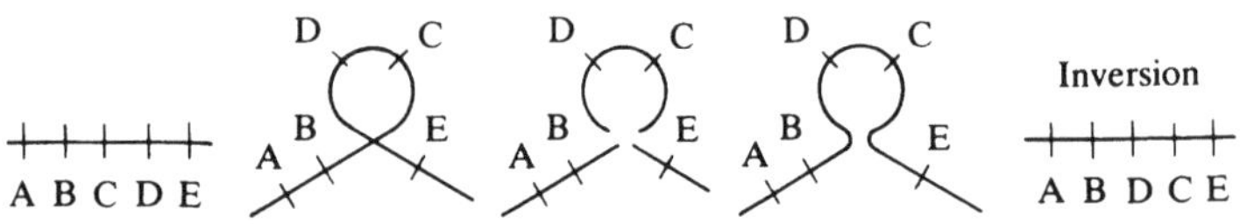

Fig. 30. Development of an inversion

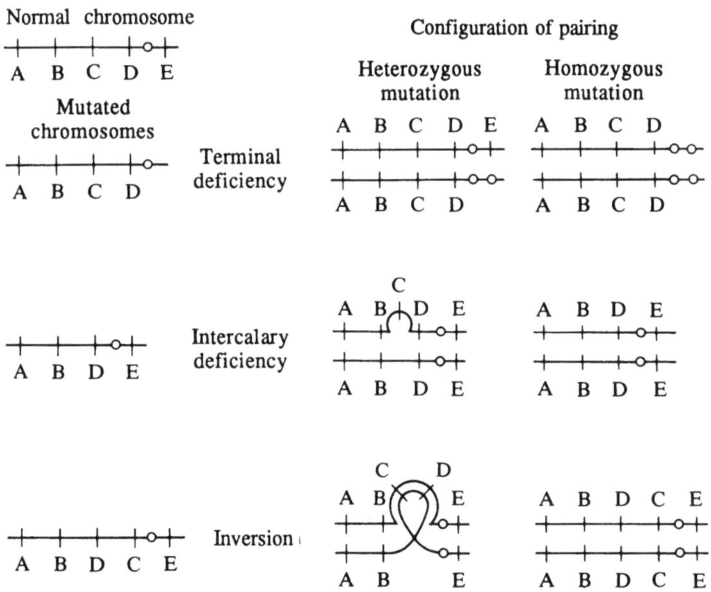

Fig. 31. Configuration of pairing chromosomes as a result of terminal and intercalary deficiencies as well as of inversions

times in populations of different organisms. The significance of inversion heterozygotes has already been pointed out in connection with the phenomena of heterosis.

If in two nonhomologous chromosomes there is one breakage each, the broken off end fragments can be exchanged by the attachment of one fragment each to the free end of fragments of the other chromosome. Such a "reciprocal translocation" initially results in a change of linkage groups (Fig. 32). At times, the genes have a different effect on the expression of characters in the rearrangement than in the original position. In this case, there is a "position effect" of genes. Plants that are heterozygous in a reciprocal translocation are characterized by semi-sterility. In the course of reduction division of a translocation hybrid, the four chromosomes are arranged in a typical cross pattern (Fig. 32), which is due to the fact that only homologous parts of two chromosomes can pair. If the four chromosomes alternately migrate to one of the poles during metaphase, the

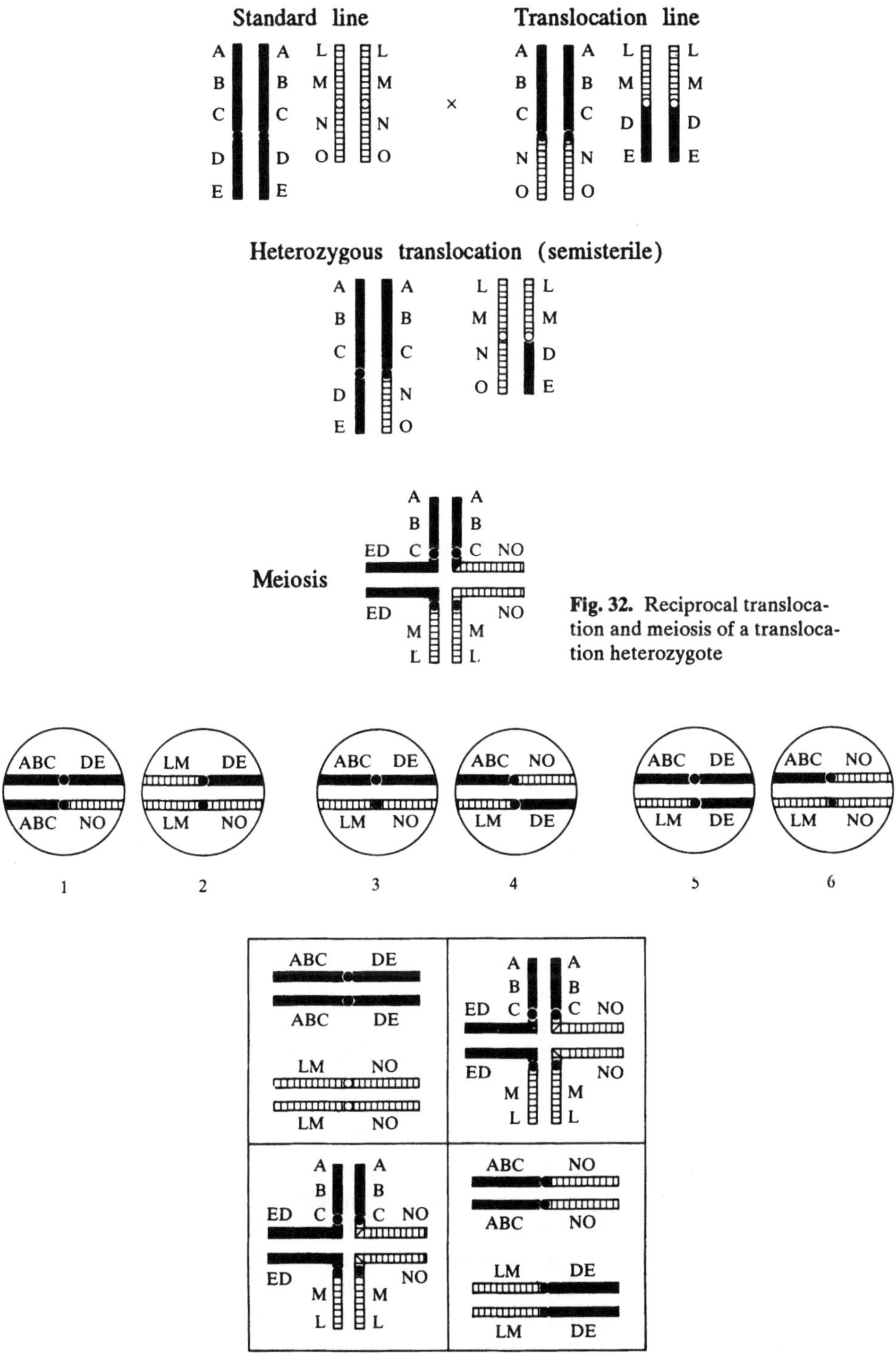

Fig. 32. Reciprocal translocation and meiosis of a translocation heterozygote

gametes are formed with the entire gene substance of both translocated chromosomes. However, if two chromosomes lying next to each other in the ring migrate to the same pole, gametes are formed that lack some gene loci, whereas other loci are present in a double dosage. These gametes are not viable. Of the six gamete types shown in Fig. 32, only gametes 3 and 4 are viable. Their combination results in three different chromosome types:

- a homozygous standard type,
- a homozygous translocation type,
- two heterozygous translocation types,

as can be seen from the combination diagram of Fig. 32. In the previous section, it has already been pointed out that by the application of mutagenic agents not only point mutations, i.e., changes at the molecular level of a gene, but also structural changes of a chromosome can be induced, which are initiated by one or several breakages within the chromosomes.

Smaller structural changes of a chromosome cannot always be clearly distinguished from point mutations in their phenotypic effects on character expression, especially if a structural change of the chromosome is not obvious by cytological investigation, and the changed phenotype shows a normal Mendelian segregation in the F_2 after a cross with the original form.

Larger structural changes, above all simple and reciprocal translocations, are relatively easy to verify in well-tested objects like barley and peas. By their cytogenetic analysis, a foundation is laid for potential application in breeding, as will be demonstrated in the following examples.

Sometimes the expression of certain characters is determined by a majority of genes that are located on different chromosomes but in the same direction, and thus they have a cumulative effect. Therefore, experimental doubling of certain genes, e.g., the gene controlling amylase production in barley, is of particular interest in breeding, if this is combined with a simultaneous increase in amylase production. Such doublings of gene loci can be induced by crossing certain translocation lines (Hagberg 1962). The diagram of Fig. 33 shows the nonhomologous chromosomes I and II of a standard line in a), which serve as original material for the subsequent experiments. The fact that the two chromosomes are not homologous can be seen by the different diagrams; their terminal points are designated by l and m or n and o, respectively. The empty circles indicate the position of the centromere. One of the chromosomes has a median, the other a submedian centromere. At the positions marked with T, breakages occur that are induced by mutagenic substances.

If there is a breakage in chromosome I at T_{Ia} and in chromosome II at T_{II}, the translocation chromosomes I_A and II_A with the terminal fragments l and n or m and o, respectively, are formed after mutual exchange of the broken off fragments (Fig. 33b). This reciprocal translocation is designated $T_{Ia\text{-}II}$; it was initially formed from the standard line as a heterozygote.

After selfings, a homozygous translocation line A can be developed. If the breakage occurs at the standard line in chromosome I at T_{Ib}, which is located at distance X from point T_{Ia} (Fig. 33a), but in chromosome II at the same position

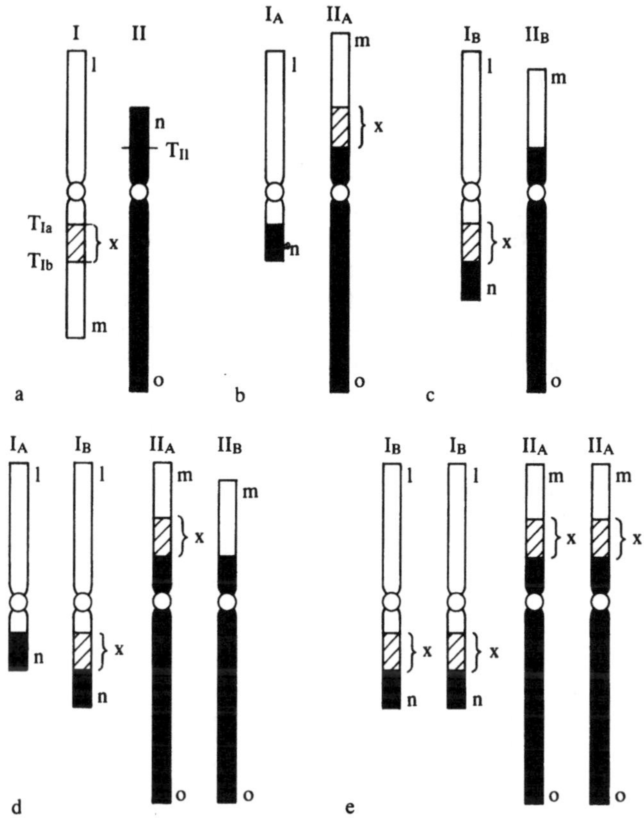

Fig. 33a-e. Development of duplicated gene loci by crossing of two translocation lines (Hagberg 1962). Chromosome structure: **a** original line (standard line); **b** translocation $T_{Ia\text{-}II}$, line A; **c** translocation $T_{Ib\text{-}II}$, line B (**b,c** gametes); **d** hybrid A × B; **e** homozygous duplication (**d,e** zygotes)

as first described in the reciprocal translocation process, then a further translocation line B can be constructed with the translocation $T_{Ib\text{-}II}$. The structures of the two rearranged chromosomes termed I_B and II_B are shown in Fig. 33c.

The two chromosomes of translocation lines A and B differ from one another in that the hatched chromosome segment X of line A is located on chromosome II_A and of line B on chromosome I_B. If these translocation lines are crossed with one another, the hybrid will have the constitution shown in Fig. 33d, with respect to chromosomes I and II; it forms the following four classes of gametes: I_AII_A, I_AII_B, I_BII_A, I_BII_B.

Assuming random pairing of these four gamete classes, 1/16 of the zygotes with constitution $I_BI_BII_AII_A$ are to be expected. As can be seen from Fig. 33e, chromosome segment X is represented in both I_B as well as in both II_A chromosomes in this plant. Thus, successful induction of a homozygous duplication has taken place by crossing two translocations comprising both identical chromosomes, which have one breakage in common in one chromosome, in the other, however, a different one. The resulting doubling of genes with equivalent effect (polymery) opens new perspectives in breeding.

In Section 2.3.2, a method developed by Ramage and Wiebe (1969) has already been pointed out, which is based on the use of experimentally induced structural changes of chromosomes. For the propagation of pollen-sterile maternal lines, Ramage and Wiebe developed a so-called balanced tertiary trisomic form. It consists of the normal diploid chromosome set and an additional chromosome (Fig. 34a). The extra chromosome originated from the translocation of chromosomes II and VII (= $T_{II\text{-}VII}$), which was induced by X-rays. The recessive gene m for pollen sterility is located in chromosome II. Chromosome VII contains the recessive gene b for white glume color. In the extra chromosome, the translocation chromosome, the dominant alleles M and B for pollen fertility and black glume color, respectively, are located. If only chromosomes II, VII, and the translocated extra chromosomes are considered, two classes of gametes are formed during meiosis of these balanced trisomes; one of these classes is characterized by the presence of the extra chromosome (Fig. 34). As egg cells, these are fully able to function. Male gametes that contain the extra chromosome, however,

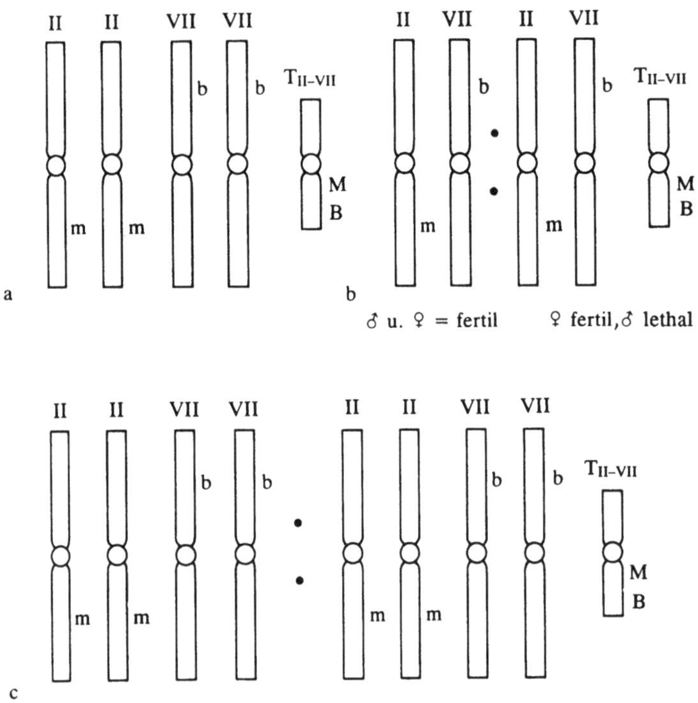

Fig. 34. ***a*** Balanced trisomic barley; ***b*** gametes of the balanced trisomic barley; ***c*** progeny of the balanced trisomic barley

die off. In the progeny of the balanced trisomes, which are pollen fertile and have black glumes, only the two genotypes shown in Fig. 34 occur: the one with genetic constitution mm bb is pollen sterile, and the other one containing the trisome is pollen fertile, because the dominant allele M is located on the extra chromosome. If these two genotypes are open-pollinated together, the pollen-sterile plants can only be pollinated by the pollen with the genetic constitution mb. This pollen is derived from the trisomes, but does not itself contain the extra chromosome, so that the embryos of the pollen-sterile plants are again of the constitution mm bb, like the female parent. All plants with white glumes must, therefore, produce pollen-sterile progeny. In the second genotype, the trisome, egg cells with the extra chromosome can be formed. These can likewise only pollinated with mb pollen. Nevertheless, plants are produced that are pollen fertile and have black glumes, because the extra chromosome contains the dominant alleles M and B.

Thus, from the glume color of the grains, the flower type, whether pollen fertile or pollen sterile, can already be predicted. B/b is a so-called marker gene enabling the breeder to perform an early selection on the desired genotype. With modern seed cleaning equipment, it is not difficult to separate the grains according to their glume color with the help of photoelectric cells. However, other marker genes that can be handled more easily may also be incorporated in this system. The shortening of the translocation chromosome, which is conspicuous in the diagram, can be explained by the deliberate induction of losses of end fragments (deletions) by X-rays. Such a fragmented chromosome is less able to form multivalents with chromosomes II and VII because of the common gene loci M and B, which would on their part lead to disorders of meiosis and consequently to the disintegration of the so-called trisome-balance mechanism.

3.1.3 Polyploid Forms

Along with an *increase* or a *decrease* in the *chromosome number*, the number of genes also changes, i.e., the entire genotype is changed. In the somatic cell of a plant, the chromosomes are present in pairs. During reduction division, the homologous partners of each pair are divided up between the two cells, thus introducing the "haplophase," during which the cells only contain one set of chromosomes; they are haploid. Fertilization by fusion of two haploid sex cells restores the "diplophase" again, the cells of which are diploid as regards chromosome number.

The chromosome number during haplophase is given as n, that during diplophase as 2n. Thus, for instance, in barley, the chromosome number of the gametes is n = 7, the number of the somatic cells, as in the embryo, the root, or the leaves, is 2n = 14.

If irregularities occur during reduction division, for example, due to nondisjunction of the partners of a chromosome pair, gametes with deviating chromosome number are formed, leading to the development of plants lacking in or having extra chromosomes. These are termed "aneuploids." If an aneuploid plant lacks a single chromosome, it is called "monosomic" (2n–1), and "nullisomic" (2n–2) if a chromosome pair is missing. In "trisomic" forms, an extra chromosome is present, in "tetrasomic" forms, one of the chromosome pairs is double. "Monoploid" or "haploid" forms represent extreme cases of aneuploidy; they only have one set of chromosomes and are subvital. Due to the lack of chromosome pairing, no sex cells capable of functioning can be produced: haploids cannot multiply sexually. Their significance in the development of homozygous inbred lines by doubling of the chromosome number has already been pointed out (Sect. 2.2.3). The phenotype of aneuploid forms strongly deviates from that of euploid original forms. Apart from a few exceptions – such as cyclamen – they do not remain constant in sexual reproduction. In vegetatively propagated species, e.g., in bulbous plants and other ornamental plants, they have achieved a certain significance in breeding.

"Polyploid" forms, which are characterized by multiplication of entire chromosome sets, are much more important in breeding than aneuploid forms. If the chromosome sets are doubled in a diploid plant, a "tetraploid" type is produced, after a further doubling, an "octoploid" one. Crosses between diploid × tetraploids initially result in "triploids," in their turn developing into "hexaploids" after doubling of their chromosome sets. "Pentaploids" result from crosses of tetraploids × hexaploids. To express the degree of polyploidy in a polyploid form, the number of chromosome sets is represented by x, while the haplophase of the diploid original form serves as a base. Correspondingly, the chromosome number of a tetraploid barley plant is given as $2n = 4x = 28$, that of a triploid as $2n = 3x = 21$, etc. If the characterization of the polyploidy level of a plant is of no interest or is not yet known, the chromosome number in these cases is given as n or 2n, depending on whether it was determined during meiosis or mitosis.

With regard to their development, polyploid species can be separated into "autopolyploids" and "allopolyploids." Autopolyploids originate from a diploid species, the chromosome sets of which have been multiplied. An autotetraploid is characterized by four homologous chromosome sets. An allopolyploid species derives from the cross between two diploid species, the hybrids of which do not pair during meiosis due to the lack of homology of the chromosome sets, and therefore do not produce any functioning sex cells; they are sterile. If the chromosome number of a sterile hybrid is doubled, the chromosomes of both species joined within the hybrid are present in pairs again, thus making it possible to produce functioning sex cells and multiply sexually. Such a fusion of two diploid species forming a third and new fertile species is termed "amphidiploidy," which is synonymous with allopolyploidy.

Auto- and allopolyploid species are fundamentally different in their cytogenetic behavior and, therefore, they also have different significance for breeding. During meiosis, autotetraploid types often form multivalents instead of bivalents, due to the presence of four homologous chromosomes each. Three to four of the homologous chromosomes form a chain or a ring, which can in turn lead to irregularities in the distribution of the chromosome sets. The resulting aneuploid gametes cause sterility symptoms and the segregation of defective forms with changed chromosome number. The Mendelian segregations of a polyploid type also differ from diploids in their numerical behavior. Allopolyploid species, on the other hand, generally act like diploid species in a cytogenetic respect. In naturally occurring species, a clear decision as to whether the are of auto- or alloploid origin cannot always definitely be made. In the course of evolution, autopolyploids can be transformed into allopolyploids forming bivalents and with constancy of chromosome sets, due to structural changes of single chromosomes. Forms that are intermediate with regard to the type of polyploidy are also called segmental allopolyploids. In breeding, point mutations and the polyploid types as chromosome mutations are the most important.

As already discussed above, all changes of the genes and chromosomes described so far are comprised in the term genome mutations; they are subject of nuclear or karyotic inheritance.

The naturally occurring polyploid types are often enlarged types of their diploid original species as regards morphology. The leaf surfaces are enlarged, as well as all flower organs. The stalks are often thickened and the height of the plants is considerably increased. This phenomenon is designated as the gigantis form. An enlarged growth type is not always based on the multiplication of the chromosome sets; in a diploid species, it may also be due to specific genes. Due to its size, the recessive heroina mutant in snapdragon (*Antirrhinum*) has the appearance of a polyploid gigantism type.

Along with the doubling of the chromosome number, physiological and ecological characters are also frequently changed. Comparative investigations of species with annual and perennial races produced the interesting result that the annual types are mainly diploid, whereas the biennial and perennial types are mainly polyploid. Thus, the annual *Urtica urens* has $n = x = 12$ chromosomes and the perennial *Urtica dioica* $n = 2x = 24$ chromosomes. Furthermore, investigations verified that diploid and polyploid races have different distribution habitats, and that often the polyploids are better adapted to extreme climatic conditions. Thus, in Schleswig-Holstein (in northern Germany), 60% of species are polyploid, in Sicily (in southern Italy), however, only 27% are.

The different distribution of diploid and polyploid types and the higher adaptability of the polyploids to extreme environmental conditions cannot be taken as a general rule. Investigations to date have shown, however, that by means of polyploidy, the genetic variability of a species is considerably expanded. Therefore, genotypes that are well adapted to extreme environmental conditions may be created to a larger extent than in diploids. This is the crucial significance of polyploidy. Autopolyploidy has contributed a great deal to the increase of diversity of types within a species. It is sometimes expressed in the arrangement of systematic subunits or in the special designation of varieties (Table 13).

Autopolyploid types can especially often be found in ornamental plants. The increase in polyploid varieties along with the intensity of breeding is a remarkable phenomenon. When *Primula malacoides*, a wild species introduced from China, was being developed as an ornamental plant at the beginning of this century, the greatest breeding successes only occurred after spontaneous formation of

Table 13. Examples for autotetraploid species

Species	Chromosome number (2n)
Zea mays	20, 30
Festuca ovina	14, 28, 42, 70
Festuca rubra	14, 42, 56, 70
Poa pratensis	42, 50, 56, 98
Phleum pratense	14, 42
Malus communis	34, 51, 68
Rubus idaeus	14, 21, 28, 42, 49, 56
Vitis vinifera	38, 57, 78

tetraploid types. The multiplicity of types originating from polyploidy and the resulting improvement in terms of habitus, number, size, and play of colors of the flowers surpass all expectations.

In the development of cultivated plants, allopolyploidy has played a decisive role. In Table 14 some cultivated plants are listed along with their diploid original species, from which they originated after a cross with subsequent chromosome doubling. Their chromosome number is equal to the sum of both of the parent species.

The spontaneous development of polyploid types can occur in two ways: first, generatively by formation of unreduced gametes. The combination of two unreduced gametes results in a tetraploid plant. The fertilization of an unreduced egg cell (2n) with a reduced pollen grain (1n) gives rise to a triploid plant (3n). In triploid plants, besides aneuploid gametes with a chromosome number between 1n and 2n, euploid 1n and 2n gametes are also formed, thus, 4n plants can also be expected in the progeny. Second, polyploidy can also originate from the

Table 14. Alloploid species and their diploid original sources

Name	2n	P_1	2n	P_2	2n
Oryza minuta	48	*Oryza officinalis*	24	*Oryza sativa*	24
Phleum pratense var. *hexaploidum*	42	*Phleum pratense*	14	*Phleum alpinum*	28
Brassica juncea	36	*B. campestris*	20	*B. nigra*	16
Brassica carinata	34	*B. oleracea*	18	*B. nigra*	16
Brassica napus	38	*B. campestris*	20	*B. oleracea*	18
Rubus maximus	42	*Rubus idaeus vulg.*	14	*Rubus caesius*	28
Prunus domestica	48	*Pr. divaricata*	16	*Pr. spinosa*	32
Aesculus carnea	80	*Aesc. hippocastanum*	40	*Aesc. pavia*	40
Gossypium barbadense	52	*Goss. thurberi*	26	*Goss. arboreum*	26
Nicotiana tabacum	48	*Nic. silvestris*	24	*Nic. tomentosiformis*	24
Nicotiana rustica	48	*Nic. paniculuta*	24	*Nic. undulata*	24
Nicotiana eastii	64	*Nic. maritima*	32	*Nic. suaveolens*	32

somatic tissue. The tissue deriving from a polyploid original cell is layer-bound; it appears at first as a mericlinal chimera, from which shoots with a uniformly modified 4n tissue layer, i.e., a periclinal chimera, can be developed (Sect. 3.2). In vegetative propagation, the periclinal chimeras remain constant. In sexual propagation, 4n offspring are only produced if tetraploidization has taken place in the L_2. Simultaneously, the periclinal structure is broken up; the offspring are tetraploid homohistonts.

For experimental induction of polyploids, it is important to know that in polyploidization within the somatic tissue, the development of constant polyploid types passes beyond a labile state, which is determined by the competition between 2n and 4n cells within a layer-bound tissue.

A systematic usage of auto- and allopolyploid types for the realization of certain breeding aims only began after 1937, when colchicine, an alkaloid from the root of meadow saffron (*Colchicum autumnale*) and a very effective agent for polyploidization, was discovered. Previously, successes had of course been achieved with other methods, but they were too complicated in their application and not effective enough to find acceptance in practical breeding.

The effect of colchicine on the dividing cells is that the spindle fiber mechanism is stopped during mitosis, so that after the chromosomes have been split lengthwise without disturbance, the chromatids do not reach the poles and consequently the subsequent cell wall formation does not take place, either. The synchronization of chromosome and cell division is interrupted. The two chromatids remain next to one another like a pair of skis in the middle of the cell, which now contains the double chromosome number.

In many cultivated plants, polyploid types were induced by the effect of colchicine, by various modifications in its application. Hereby, a variation of the following factors is especially important: concentration of the agent, duration, and timing of treatment. In general, solutions of 0.1–0.5% are used for a period of 6–24 h. A weak solution combined with a longer duration of treatment is to be preferred. Solutions that are too strong have a toxic effect or, due to the continuing effect of colchicine and the resulting failure of the spindle fiber mechanism, highly polyploid cells occur after further divisions. The longer the treatment with a weaker solution, the greater is the chance that the colchicine solution can have an effect on dividing cells.

Cell divisions take place in the meristematic tissue of various organs, as in germinating seeds, in parts of the shoots, and in flowers, so that these organs are more or less suitable for colchicine treatment. The widespread treatment of seeds by immersing them in a solution is quite simple, of course, but considerable losses occur due to root damage if these are not protected from the detrimental effects of the colchicine. By application of certain modified methods of seed treatment, the yield of tetraploid plants can substantially be increased. If the top layers of sand in a ramet is covered with a layer of agar after the sprouting of the radicle and a colchicine solution is applied onto this, the roots are completely protected against the colchicine. Another effective method is to remove the seed kernel including the scutellum, from which the seedling obtains its nutrition, from the young seedling plants. After cutting back the first cotyledon to 2 cm, a drop of colchicine-locoweed (*Astralagus*) slime is smeared on the seed site. Plastic foil is then wrapped around the young seedling to prevent the penetration of colchicine into the original substrate. Covering the ramets with glass prevents the carrier substance of the 0.25% colchicine solution from drying out too quickly. Application of colchicine can also take place by injecting a solution into the developing shoot of a seedling or by inverted immersion of cut seedlings into a solution. A quite simple method is to lay cotton pads soaked with a colchicine solution onto the axil of developing side shoots; its usage is preferred in species with easily accessible vegetative cones, such as rape. A repeated replacement of the soaked cotton pad is necessary. A quite original method was developed in onion, which had resisted successful polyploidization for a long time. By removal of the flower buds, the receptacle is stimulated to form bulbils. If the receptacle is treated with a colchicine solution after removing the flower buds, tetraploid initial cells are formed among others, from which tetraploid bulbils are developed. A disadvantage of this method is that success cannot be determined until 3 years later, after the normally formed onions have set seed.

The plants developing after a colchicine treatment are designated as C_1 generation, sometimes also C_0. In this generation, selection on polyploid types begins. First, the normal plants, which cannot be distinguished from the untreated control, are removed. Among the remaining plants, only those are subjected to special observation which are just far enough modified in form and development that normal multiplication can be expected. Changes of the leaf-area-index and the leaf color, thickening of the stalk, and enlargement of the flower organs are signs of a successful polyploidization. Comparative tests give first indications of the size of the guard cells of the somatas, number of chloroplasts, and size of the pollen grains.

Many of the C_1 plants that are regarded as polyploid are chimeras, so not infrequently normal side shoots develop. If these are not pruned, a complete normalization can occur due to the suppression of the polyploid sector. Under optimum culture conditions, the 4x sectors may be favored in their competition with 2x sectors. For the same reason, the seed progenies segregate into diploid, polyploid, and aneuploid plants, so that another selection must take place in the C_2 generation, if possible by ascertainment of the chromosome number in roots or shoots.

In the last four decades, a large number of polyploid types has been produced in very many species, the majority of whose breeding value cannot be perceived completely. Often, the new crude polyploids display certain gigantism characters with regard to size of cells, leaves, and flowers, of shoot length and seed weight. Distinct rules of the relationship between polyploidy level and a certain character change do not exist, however. Only the cells are larger in all polyploids investigated thus far; the extent of cell enlargement can vary according to individual species or even according to different varieties of the same species: the cell enlargement index is determined by genome and plasmon. Even if the cell size is of fundamental importance for the formation and size of the organs, all organs still also have their own independent genotypic basis. Two tetraploid races of *Impatiens balsamina*, for example, display completely contrasting behavior: one of them has longer shoots and a higher fresh and dry weight than the diploid original type, while the other one is characterized by lower values of shoot length, fresh, and dry weight.

The same situation exists in the comparison of physiological characters between diploids and tetraploids. In general, a higher water content has been found in tetraploids so far, diluting the cell constituents and leading to lower osmotic values. The physiological characters of a cell, which simultaneously change with cell size, result in extensive changes in the total development of the organism and are also, therefore, one of the most important causes for the frequently decreased fertility of experimentally induced polyploids. Fertility is not only determined by the processes connected with the changed cell size, but is further based on the functioning of a delicately balanced pollen tube stigma system, in which the individual morphological parts and osmotic relationships are again genotypically controlled and fixed. When comparing diploids and tetraploids, close attention must also be paid to the external conditions, since the individual polyploidy levels are not always changed in the same direction. The differences are frequently much more prominent under unfavorable than under favorable culture conditions. Especially with regard to physiologically caused fertility disorders of crude polyploids, the particular requirements in the cultivation of polyploid types must be taken into account by corresponding environmental conditions. To understand the different behavior of naturally occurring and experimentally induced polyploids, a brief digression on the genetic behavior of autotetraploid hybrids is necessary. The knowledge thus obtained also contributes to the development of breeding processes for further treatment of crude polyploids.

If a gene behaving according to Mendel's laws with the alleles A and a is assumed, the two homozygous genotypes AA and aa and one heterozygous genotype Aa can be realized in the diploid type. The corresponding autotetraploid type can also produce only two homozygous genotypes AAAA and aaaa, but three different heterozygotes AAAa, AAaa, and Aaaa. Simplified, these five genotypes can be designated as follows:

A^4, A^3a^1, A^2a^2, A^1a^3, and a^4.

The three heterozygotes segregate differently in the F_2. During meiosis, two of four homologous chromosomes migrate to one pole each. In which way the combination of two chromosomes arises in an A^3a^1 genotype can easily be gathered from the following pattern, in which the homologous chromosomes are designated 1–4:

Homologous Chromosomes (1–4)	1 A	2 A	3 A	4 a
1 A	–	AA	AA	Aa
2 A	–	–	AA	Aa
3 A	–	–	–	Aa
4 a	–	–	–	–

Thus, AA and Aa gametes are produced in the same ratio. Their combination results in the three genotypes A^4, A^3a^1, and A^2a^2 in a ratio of 1:2:1, as can be seen in the subsequent pattern:

Gametes	AA	Aa
AA	AAAA	AAAa
Aa	AAAa	AAaa

If complete dominance of A over a is assumed, there is no phenotypic segregation in the F_2 of an A^3a^1 hybrid. That a new heterozygote A^2a^2 is formed by recombination can only be recognized in the (F_3) progeny of F_2 plants. This A^2a^2 heterozygote forms 1 AA : 4 Aa : 1 aa gametes, the combination of which results in 1 A^4 : 8 A^3a^1 : 18 A^2a^2 : 8 A^1a^3 : 1 a^4 genotypes, as shown in the following patterns:

Gametes	1 AA	4 Aa	1 aa
1 AA	1 AAAA	4 AAAa	1 AAaa
4 Aa	4 AAAa	16 AAaa	4 Aaaa
1 aa	1 AAaa	4 Aaaa	1 aaaa

If the A allele is completely dominant over all a alleles, a phenotypic segregation of 35:1 occurs. Within this F_2, the third heterozygote A^1a^3 occurs. It forms 1 Aa : 1 aa gametes, the combination of which gives rise to 1 A^2a^2 : 2 A^1a^3 : 1 a^4 zygotes, as can be ascertained by application of the patterns already used above. There is a phenotypic segregation ratio of 3:1, differing from a diploid 3:1 segregation, however, in that it contains only *one* homozygous genotype, but two different heterozygous genotypes.

Assuming complete dominance of one A allele over all a alleles as in the above-mentioned example is only one of the possible relationships between the alleles of a tetraploid gene locus. In cyclamen, the F^4 genotype has colorless flowers. With an increase in the number of f alleles, the red color is intensified: F^3f^1 light red, F^2f^2 pink, and F^1f^3 red. This is a so-called dosage effect of alleles, a phenomenon that offers great advantages in breeding polyploids. In dahlia, gene I, which is responsible for the production of ivory-colored flavones, already reaches its maximum effectiveness with two dominant alleles, so that the genotypes I^2i^2, I^3i^1, and I^4 cannot be phenotypically distinguished from one another.

Considering such different behaviors of single gene loci at the tetraploid level, it is understandable that the genetic system adjusted to the diploid level no longer functions after tetraploidization. Only after recombination of genes and the simultaneous beginning of selection can genotypes be developed in which the interaction of genome and plasmon in general and the interaction between the individual gene loci (also between the alleles of each individual gene locus) correspond so well with each other, that a harmonious development is guaranteed and, with it, adaptation and performance of the new tetraploids.

How the performance of a tetraploid is determined by the genetic constitution of the diploid original type and how the value of a crude tetraploid can be increased by recombination and selection is explained by the following model (Table 15). Gene F increases resistance to frost by 1 °C, inhibitor gene H compensates for the effect of F again; recessive alleles h and f are ineffective. The frost hardiness of the four diploid genotypes FFHH, FFhh, FFHh, and FfHh can be seen in the corresponding column of the table. After tetraploidization of FFhh and FFHh, the value is raised due to an increase in frost resistance, while the tetraploids deriving from the genotypes FFHH and FfHh show no change in this character. A selection in the following generations of F^4H^4 remains unsuccessful, whereas in $F^2f^2H^2h^2$, frost hardiness can be increased from 0 to –4 °C.

This model clearly demonstrates that polyploidy breeding requires intensive treatment of the crude polyploids for several generations and this in turn assumes original material that is as heterogenic as possible. During the breeding process, a new heterozygous population can be developed for further selection by crossing initially selected tetraploid types with valuable characters.

Comparative performance tests over several years and, above all, at ecologically very different locations are necessary to learn about their adaptability and efficiency under different environmental conditions, and to ascertain their special requirements for cultivation techniques with regard to sowing time and sowing density and the economical utilization of increasing fertilizer applications. Tetraploid varieties are not enlarged versions of their diploid original species, but are sometimes new species of cultivated plants in terms of their ecological requirements, their physiological behavior and, consequently, their economic performance.

Table 15. Effect of selection in autotetraploids (After Melchers 1946)

Hypothesis: F increases frost resistance by 1 °C
H compensates the effect of F

Diploids Genotype	Frost hardiness (°C)	Autotetraploids Genotype	Frost hardiness (°C)	Effect of selection in subsequent generations
FFHH	0	FFFFHHHH	0	None
FFhh	–2	FFFFhhhh	–4	None
FFHh	–1	FFFFHHhh	–2	–4 °C: FFFFhhhh
FfHh	0	FFffHHhh	0	–4 °C: FFFFhhhh

The proportion of tetraploid varieties in many cultivated species has steadily increased in the recent past. At the same time, maintenance breeding and further improvement of 4n varieties must meet new requirements. An important problem is improvement of fertility, especially by stabilizing the tetraploid condition. Formation of multivalents during meiosis and resulting irregularities in the distribution of homologous chromosomes, which leads to the formation of aneuploid gametes and consequently to aneuploid plants, have already been pointed out above. This downward adjustment can be prevented by selection of genotypes in which a favored bivalent formation is genetically conditioned, or by automatic bivalent formation due to spontaneous or induced structural changes of two or four homologous chromosomes. These processes are designated by the term "diploidization," which is not very accurate. They are rather allotetraploidizations and the more sutiable term would be "allo-" or "amphidiploidization." Another cause for reducing the level of ploidy of tetraploids with neighboring diploid varieties. In the progenies of these triploids, mainly aneuploid plants are formed. Since aneuploid plants can be identified by their weaker growth, they should again be removed during the multiplication steps to prevent further contamination. This purification is facilitated by wide spacing of the plants.

That breeding to constancy of certain characters or their maintenance within the varieties is hindered by the tetrasomic segregations can be explained by the behavior of the progeny of the three different heterozygotes already described above, and by the different dosage effects of the four alleles of each gene locus. Further improvement of tetraploid varieties is achieved by combination breeding. The tetrasomic segregations make practical application of combination breeding even more difficult than breeding by selection. In monohybrid segregations and dominance, 1/36 of recessive types are to be expected, in a dihybrid segregation, on the other hand, only 1/36 x 1/36 = 1/1296. This by no means implies, however, that a particularly extensive material must be cultivated in the F_2 to realize the desired gene combinations in this generation. By cultivating many individual progenies and later generations, a segregation of the recessive characters can be expected even if the number of individuals per progeny is rather small. Thus, one must attempt to attain the breeding goal in steps. The question arises as to whether or not in certain cases induced reduction of the chromosome number by half, i.e., generation of dihaploids, could possibly facilitate combination breeding in tetraploids. The desired diploid combination would then have to be tetraploidized again. The great advantages of polyploidy breeding are primarily based on the larger genetic variation of the polyploids compared to that of diploids and to the more numerous combination possibilities connected with this.

The enlargement of the organs (the gigantism character) and the resulting increase in performance are only two of the many consequences of the increased variation after polyploidization. Occassionally, a desired increase in performance is already optimally achieved on the triploid level. Thus, for instance, triploid alders are characterized by particular vigor, whereas tetraploid alders are less vigorous. In vegetatively propagated species such as alders, poplars, apples, bananas, sugarcane, tulips, and others, triploid hybrids can be bred without difficulty. In sexually propagated cultivated plants, the standard seed must be continuously produced anew by crossing a diploid and a tetraploid parent line, i.e., the breeder thus enters at the same time the domain of hybrid breeding, the methods of which have been described in detail previously in Section 2.3.

The breeding methods overlap a great deal, as shall briefly be demonstrated by the development and use of triploid watermelons. A great disadvantage of watermelons, which are especially preferred in the warm countries, is that they contain so many seeds. The triploid hybrids from the cross of diploid varieties with experimentally induced tetraploids are seedless; triploids produce neither egg cells nor pollen grains capable of functioning. Since prickle pollination is necessary for the development of a fruit, some diploid plants must be added to a stand of triploids, the pollen of which induces fruit development of the triploids, without forming a viable embryo, so that the plants remain seedless. A prerequisite for the development of a usable triploid hybrid variety is the testing of a multiplicity of diploid and tetraploid lines for their combining ability.

It must be qualified that, for reasons of reproductive biology, triploid varieties cannot be produced in all species.

The allopolyploid character of a cultivated plant can only be inferred by careful cytogenetic analyses, i.e., by crosses between the two assumed original species, from which cross the new amphidiploid species is derived after subsequent chromosome doubling. Thus, for example, rapeseed (*Brassica napus*, 2n = 38) originated from the cross between cabbage (*Brassica oleracea*, 2n = 18) and turnip rape (*Brassica campestris*, 2n = 20). If rapeseed is crossed with cabbage, nine bivalents and ten univalents are formed during the meiosis of the hybrids; in the hybrid rapeseed × turnip rape, however, ten bivalents and nine univalents can be observed. If, based on such analyses, the diploid original types are known, a resynthesis of already existing allopolyploid species can be carried out. The breeding value of an allopolyploid species is largely determined by the genotype of the diploid original species. Prior to resynthesis, a selection among the genotypes of the species to be crossed must therefore take place. Since the outcome of the combination in terms of its breeding value cannot be safely predicted, a large number of different genotypes should be used for the cross. Reciprocal crosses are also recommended, since interactions between plasmon and genome can frequently be observed after interspecific crosses.

Resynthesis may be carried out in two ways: by polyploidization of the sterile F_1 hybrid or by initial tetraploidization of the diploid original species; their subsequent cross results in amphidiploid hybrids that are already fertile. This tetraploidization of the parent plants is especially advisable if the crosses are difficult to carry out, if only little F_1 seed for polyploidization is available, or if polyploidization of vegetative parts of the hybrids is not promising.

Resynthesized amphidiploid cultivated species largely correspond to the natural amphidiploids in morphological and physiological respect and, after crosses, they generally result in fully fertile hybrids. In practical breeding, they have also already been successfully applied, e.g., in rape.

A productive field for amphidiploidization is offered by ornamental plant breeding. In some genera, e.g., begonia and rhododendron, there are interspecific hybrids that can only be vegetatively propagated, because they are sterile. Besides being more costly in economic terms than seed cultivation, vegetative propagation also has the disadvantage that infectious diseases – particularly virus diseases – are transferred vegetatively. By polyploidization of such sterile hybrids it is possible to grow these varieties true to seed and, consequently, to lower their cultivation costs. Furthermore, they can be used in combination breeding, thus enlarging and improving the range of corresponding cultivated species.

If the parent plants of sterile ornamental plant species are known, the possibility of resynthesis should also be weighed; by the use of different genotypes as parent plants, diversity of corresponding allopolyploid species can simultaneously be increased. Finally, synthesis of new amphidiploids not yet present in nature and their development into new cultivated species may offer a field of breeding activity for the future that has not been much exploited thus far. In this context, amphidiploids from crosses between *Triticum* and *Aegilops*, *Triticum* and *Agropyrum*, *Triticum* and *Secale*, *Raphanus* and *Brassica*, and others are called to mind. Often, these are crosses between a cultivated species and a wild species.

3.1.4 Variability in Cell and Tissue Culture

Since the 1970s, cell and tissue culture procedures have been widely used in horticulture and agriculture; the simplest method is the rapid propagation of a plant from parts of its tissue. The success of this technique is shown, e.g., by the inexpensive orchids available from Singapore. Many ornamental plants now come from material that has been propagated many hundred times over in the test tube. In such in vitro cultures, which are kept – in contrast to field-grown in vivo cultures – under axenic conditions, quite often a new genetic variability has been observed. As long as the material does not become completely dedifferentiated – as in cultures starting from axillary buds, shoot tips, or meristems – the genetic stability is quite high. If, however, the initial material used is further reduced in size and one gets down to individual cells or, if the solid cell wall is removed, to protoplasts, plants regenerated from such cells – normally via an unorganized callus phase – show genetic variability, the amount of which increases with the time of undifferentiated growth.

For regeneration of tissues or cells in vitro, the donor material must be made axenic, e.g., by aqueous NaOCl solutions (1–3% NaOCl). Such aseptic tissues are placed on solid or liquid culture media that contain macro- and micronutrients, i.e., anorganic salts, a carbohydrate, phytohormones, and vitamins. One standard medium is given in Table 16. For solidifying the medium, either natural agar or artificial gellants like gelrite are used, in concentrations between 0.1 and 1%, but also cheap media solidified with starch can be used. In liquid media it is possible to culture cells also as suspensions and to produce, e.g., secondary compounds. If the composition of the culture medium is suitable, tissue and cells start to proliferate and form either an unorganized cluster of embryo-like structures. The embryo will form a plantlet directly, while the callus normally starts morphogen-

Table 16. Murashige and Skoog medium (1962) and Linsmaier and Skoog medium (1965)

Macro nutrients	mg/l	mM	Minor nutrients	mg/l	µM
KNO_3	1 900	18.8	H_3BO_3	6.2	100.0
KH_2PO_4	170	1.3	$MnSO_4 \cdot 4\,H_2O$	22.3	100.0
NH_4NO_3	1 650	20.6	$FeSO_4 \cdot 7\,H_2O$	27.8	100.0
$MgSO_4 \cdot 7\,H_2O$	370	1.5	$Na_2EDTA \cdot 2\,H_2O$	37.3	100.0
$CaCl_2 \cdot 2\,H_2O$	440	3.0	$CoCl_2 \cdot 6\,H_2O$	0.025	0.15
			$CuSo_4 \cdot 5\,H_2O$	0.025	0.1
			$ZnSO_4 \cdot 7\,H_2O$	8.6	30.0
			$Na_2MoO_4 \cdot 2\,H_2O$	0.25	1.0
			KJ	0.83	5.0

Ion composition	mM		µM
Na^+	0.2	Mn^{++}	100
K^+	20.1	Fe^{++}	100
NH_4^+	20.6	Co^{++}	0.15
Mg^{++}	1.5	Cu^{++}	0.1
Ca^{++}	3.0	Zn^{++}	30.0
SO_4^{--}	1.7	BO_3^{---}	100
NO_3^-	39.4	MoO_4^{--}	1.0
PO_4^{---}	1.3	J^-	5.0
Cl^-	6.0		

Supplements to Murashige and Skoog medium

	mg/l	mM		mg/l	µM
Edamin	1 000	–	Glycine	2.0	26.7
Myo-inositol	100	0.6	Thiamine HCl	0.1	0.3
Sucrose	30 000	87.8	Pyridoxine HCl	0.5	2.5
			Nicotinic acid	0.5	4.1
			IAA	2.0	11.4
pH 5.7–5.8			Kinetin	0.2	0.9

Supplements to Linsmaier and Skoog medium

	mg/l	mM		mg/l	µM
Myo-inositol	100	0.6	Thiamine HCl	0.4	1.2
Sucrose	30 000	87.8	IAA	2.0	11.4
pH 5.6			Kinetin	0.2	0.9

esis via organ formation (organogenesis). The faster an organized plant is formed, the lower is the rate of in vitro mutations. But even when no callus is apparent, such variability is found. Most frequently, aneuploid genomes and polyploids are found that are only of limited interest to applied plant breeding. Although the reasons for the changes are not understood at all, there are some hypothetical explanations, for example, that under the in vitro culture stress, mechanisms are activated which cause a reorganization of the genomes, or that mobile elements (McClintock 1950) may be responsible. An understanding of why the procedure works is surely not a prerequisite for using it.

Larkin and Scowcroft (1981) called the variability from callus "somaclonal variation." If protoplasts are the starting material, it is termed "protoclonal variation," and if originating from gametes, it is "gametoclonal variation." In all instances, the variation is a specific form of mutagenesis. The in vitro culture stress coupled with processes of de- and redifferentiation act as a mutagen. Thus, from a given homogeneous cell population (homo- and heterozygous), new phenotypes are expected that occur without any meiotic recombination. Similar to the expectations in classical mutation breeding, it was hoped to find useful somaclonal variants among plant populations regenerated from in vitro cultures. Larkin and Scowcroft (1981) judged this variability to be a useful technique for intracultivar improvement, and described several mutants in their review that were supposed to be useful. Later it was demonstrated in tobacco that such characters induced in vitro showed Mendelian segregation. Somaclonal variation has gained some importance in breeding for disease resistance. In addition to disease resistance, somaclones are used for detecting herbicide tolerance.

It soon became obvious that experiments making use of somaclonal variation should start from single cells rather than from larger pieces of tissue or callus, for somaclonal variation to be superior to classical mutation breeding. A system that starts with single cells can help circumvent problems of cross-feeding and chimerism, thereby increasing the chance of finding valuable mutants. Furthermore, the number of individuals can easily be increased to 10^6 in vitro, which is within the range of the natural mutation rate of 10^{-6}. Also, the use of haploids is advantageous, as recessive mutants are visible. Of particular interest are selection systems detecting desired genotypes, applicable in vitro, to enrich the populations. These are possible advantages compared to classical mutation breeding procedures.

It cannot be overemphasized, however, that there is no general need to increase the genetic variability artificially in vitro. Sufficient variability already exists or can be found by using sexual recombination in most crops. It is only for certain specific traits, lacking in nature, or for certain specific plants such as trees with a long regeneration cycle, that somaclones may be preferable. This means that for in vitro culture there is generally a need to prevent, rather than to increase, somaclonal variation. Currently, the best procedure for stabilizing tissue cultures is to subject them to a rapid passage through the unorganized callus phase. In vitro conditions must be a compromise between a phytohormone-rich medium that will initiate growth (e.g., one high in 2,4-D but which causes additional alterations) and a phytohormone-poor medium, which keeps the genotypes stable and prevents the survival of too many aberrants.

The progeny of several somaclones resistant to toxins of pathogens quite often expressed significant differences compared to the level of resistance of the starting material. In most instances, such differences were not significant, however, when populations regenerated from somaclones were tested under field conditions. In barley, potato, and wheat, the results of 2–3 years of progeny tests using natural or artificial infection on plants, preselected in vitro by embryogenic suspension cultures of protoplasts, showed phenotypic differences in the level of resistance towards both the negative and the positive side. This was in contrast to the reaction of the first in-vitro-derived generation, which showed significantly improved somaclones. It is probably important to use, for the in vitro screening, selective agents that are essentially correlated to the infection process, e.g., the toxin Sirodesmin PL in the case of *Phoma* selection in *Brassica*. In any case, the genotypic differences following in vitro mutagenesis are smaller than the differences used by the breeder after meiotic segregation. It is still under discussion whether this lack of significance is due to the absence of genetic difference or whether the test procedures are not adequate to detect and ensure slight, probably quantitative, alterations. One must know more about the actions of the pathogen or the reactions of the infected plant to develop precise strategies that will make possible the detection of useful somaclones.

3.2 Use of Periclinal Chimeras

The formation of somatic mutations, so-called sports, and the versatile possibilities of their usage in breeding can only be understood if the structure of the apex of the shoot is known. All organs of a plant originate in the apex. In flowering plants, the developing vegetative cone consists of three different tissue layers, termed L_1 (dermatogen), L_2 (subdermatogen), and L_3 (corpus). From the L_1 the epidermis is formed, from the L_2, e.g., the mesophyll and the gametes, and from the L_3, e.g., the vascular bundle, the flesh, the pith, and the adventitious roots. Besides the L_3 layer, the L_2 layer sometimes partially participates in the organization of the inner corpus. Usually the cell division in L_1 and L_2 is anticline, i.e., vertical to the layers, so that these layers show surface expansion only, whereas in the L_3 there are anticline and pericline divisions. Consequently, the cells also divide parallel to the layers.

If a mutation occurs in a cell, it is more or less bound to the layer. The mutated cell can expand by anticline divisions within its layer only, resulting first in the formation of a mericlinal chimera, i.e., a tissue layer consisting of normal and mutated cells as illustrated in the upper part of Fig. 35 (the mutated part of the cell layer is printed in boldface). If from such a mericlinal vegetative cone a shoot develops, sectors with an altered feature can be formed, as shown in the lower part of Fig. 35. An unstable mericlinal chimera can be seen in Fig. 35b. A tissue layer comprising mutated cells only, e.g., an epidermis, can participate in the formation of a lateral shoot; the mutated L_1 layer then covers the genetically unchanged L_2 and L_3 layers like a coat. A stable mericlinal chimera (Fig. 35c) is then formed, a so-called periclinal chimera, which is characterized by tissue layers genetically differing from one another. If the lateral shoots are formed without the involvement of tissues of the mutated sector, the mutation in this spot is lost (Fig. 35d).

The mutation can be of the genic or of the chromosomal type. Since vegetatively propagated plants are, in general, highly heterozygous and gene mutations mainly change an allele from a dominant to a recessive state, the mutation will manifest itself as a recessively conditioned new formation in the morphological or physiological behaviour of the whole plant as periclinal chimera, provided that the original periclinal chimera covers the entire layer affected by the mutation. The periclinal chimeras mostly remain stable when propagated vegetatively; in the case of sexual propagation, the genetic behavior of the progeny is solely determined by the genetic constitution of the L_2 (Sect. 3.1). The layer-bound gene mutations can only manifest themselves, however, if the features controlled by the mutated genes are also formed from the layer in which the mutation takes place. This means that in vegetatively propagated plants, many mutations can be present as nondetectable cryptic mutations. The older a clone, the more cryptomutations it will contain; examples of this are the old wine varieties.

The somatic mutations can be expressed by conspicuous morphological changes such as the color of the flowers, lack of thorns, and others, or even by small changes of quantitative features that can often only be observed as such if the clones are examined for differences. Periclinal chimeras are not only characterized by different genetic constitutions of the single tissue layers, but sometimes

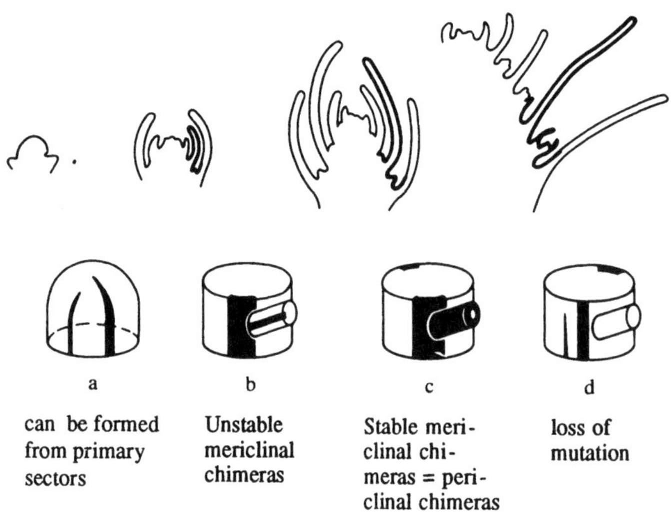

Fig. 35. Periclinal chimeras. Growth of the mutated tissue (Nybom 1961)

also by different chromosomal constitutions. The apple variety "Winesap", for example, has a diploid L_1 and L_2 layer, whereas the L_3 is tetraploid. Since the L_2 and L_3 tissues participate in the formation of the fruit pulp, the apples are distinguished by an asymmetrical gigas shape.

Occasionally there are spontaneous rearrangements of the tissue layers in periclinal chimeras. Since these can be experimentally induced, they are important aids in the analysis of periclinal chimeras. The rearrangement of genetically differing layers often leads to new phenotypes, mostly by activation of cryptic mutations, and therefore increases the variation of vegetatively propagated species.

A rearrangement of the layers can take place in such a way that, e.g., cells of the L_1 migrate into the L_2 layer by periclinal division or, conversely, cells of the L_2 can break through the L_1 layer. These processes can often be observed if, through external influences – such as X-rays or other mechanical changes – one or the other layer has been destroyed. The induction of genuine somatic mutations using x-rays is dealt with in Section 3.1. Different possibilities of layer rearrangement are illustrated in Fig. 36. In Fig. 36a, a periclinal chimera is shown with a mutated L_1, a so-called ecto-chimera, changing into a diecto-chimera, which is a periclinal chimera with a mutated L_1 and L_2 layer. In Fig. 36b there is a meso-chimera, a periclinal chimera with a mutated L_2, where the L_1 as well as the L_3 are replaced by the mutated L_2, so that from a periclinal chimera, a mutant possessing genetically uniform tissue layers is formed. A similar process is taking place in Fig. 36c, differing only in that a mutated L_3 of a so-called endo-chimera is replaced by a normal L_2, thereby restoring the normal original form.

Due to their genetically differing layers, periclinal chimeras are also termed heterohistonts. From them, homohistonts can be obtained by induction of lateral

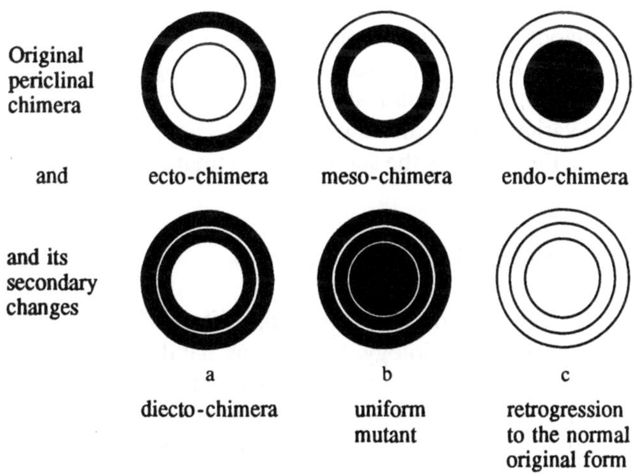

Fig. 36. Different types of periclinal chimeras (Nybom 1961)

buds out of the L_3 layer and by root propagation. For this, it is necessary to decapitate a shoot and to eliminate all lateral buds, which stimulates the formation of side buds from the phloem (L_3). After cutting out all eyes of potato tubers, adventitious shoots regenerate out of the inner tissues after 4–8 weeks. Using this method, Klopfer (1965) was able to prove that the potato variety "Rote Erstling" originating from "Erstling" is a monoecto-chimera. After removing the buds of red tubers the regenerates resulted in the original white-tubered "Erstling". The heterohistont has reorganized itself into a homohistont again. A further proof for Rote Erstling being a monoecto-chimera could be obtained by crossing "Rote × Weisse Erstling". All progeny produced white tubers, i.e., the L_2 layer of "Rote Erstling" had remained unchanged; it was genetically the same as that of "Erstling". The sexual transfer of valuable features from monoecto-chimeras is therefore dependent on a preceding change into a meso-chimera.

In apple varieties with a tetraploid L_3 layer, pure tetraploid homohistonts can be obtained by induction of adventitious buds, which also produce a tetraploid progeny after sexual propagation.

The induction and use of periclinal chimeras, especially by methodical induction of layer rearrangements, is of interest in plant breeding for a variety of reasons; it has a potential of breeding possibilities that has not yet been fully exploited, especially in ornamental plants, as has been shown in experiments with *Euphorbia*, *Pelargonium*, *Dahlia*, *Rosa*, and others.

Today, by application of methods developed in recent years for the regeneration of plants from parts of tissues that can be grown in culture media, it is possible to obtain uniform mutants from chimeras much more systematically.

Using somatic mutations in single tissue layers, additional desired changes in certain traits can be induced in valuable varieties, without losing the valuable basic character of the original variety.

The fact that many vegetatively propagated plants are periclinal chimeras to a much higher degree than imagined earlier, and that latent genetic variability can develop by means of layer rearrangement can explain the breeding success of a clonal selection, e.g., in potatoes and vines, which had not been understood before. Another proof for the relatively high genetic variability of old vegetatively propagated, cultivated varieties lies in the existence of various clones of a vine variety, which are characterized by adaptation to special locations. Furthermore, this explains the success of systematic clonal selection.

3.3 Selection Breeding in Apomicts

Development and application of the different breeding methods strongly depend on reproductive biology. On the one hand, it is important whether or not the cultivated species is propagated generatively with seeds and whether the seed development here results from self- or cross-fertilization and, on the other hand, whether or not vegetative propagation with tubers, bulbils, rhizomes, scions, cuttings, etc. is possible.

In generative propagation, two gametes fuse after meiosis of the macrospore and microspore mother cells and form egg cells and pollen with only one set each of chromosomes, whereas the somatic cells have two chromosome sets, i.e., the number of chromosomes is double. This kind of multiplication is called *amphimixis*, which is the opposite of *apomixis*, a multiplication in which no fusion (mixis) of two gametes takes place. In a broad sense, apomixis includes vegetative propagation using tubers, bulbils, rhizomes, and viviparity, whereas in a narrow sense, it is the so-called agamospermy, i.e., seed formation without fusion of two reduced gametes. This chapter deals with the latter in regard to the consequences for the breeder, and will be referred to as apomixis, as it has come to be known.

Before considering the nature of apomixis, the normal embryo sac development, occurring in many variations, will first be schematically depicted for comparison purposes (Fig. 37). The diploid nucleus of the embryo sac mother cell, no. 1), undergoes meiosis in two division steps. During the first division, the number of chromosomes is reduced; the second step is a mitotic division resulting in a tetrad with four haploid macrospores. Three macrospores of this tetrad degenerate as shown in no. 2), while the remaining one differentiates into an embryo sac after three further divisions of its haploid nucleus into eight nuclei. These are grouped as follows: the egg cell at one pole with one synergid on each side, on the opposite pole three antipodes, and in the middle close to one another, the two pole nuclei, which fuse to form the secondary embryo sac nucleus.

The pollen tube growing on the stigma releases two generative nuclei of the pollen into the embryo sac. One fuses with the egg cell, which now forms the embryo, whereas the fusion of the second nucleus with the two pole nuclei initiates the development of the triploid (containing three sets of chromosomes) endosperm.

In *apospory*, a form of apomictic multiplication, a meiosis begins first in the embryo sac mother cell, but stops already during the first division or in one of the following steps. Instead of the now degenerated embryo sac mother cell, a diploid embryo sac develops in a somatic cell in the vicinity of the embryo mother cell, as is schematically shown in no. 3) of Fig. 37. The diploid, i.e., nonreduced nucleus of this somatic cell, undergoes three further divisions, so that the apospore embryo sac contains eight diploid nuclei, one of which functions as diploid egg cell.

In *diplospory*, which can also be regarded as apomictic multiplication, the embryo mother cell undergoes meiosis, however, by the reunification of two reduced nuclei, a diploid nucleus is again formed. By means of three division steps of this so-called restitution nucleus, a normal embryo sac is formed with the eight diploid nuclei.

From the diploid egg cells, direct parthenogenetic development of embryos, i.e., without fertilization, is possible in apospory as well as in diplospory. If the diploid egg cell is fertilized, a triploid zygote is formed. Occasionally, however, secondary fertilization of the two pole nuclei is necessary for the development of the endosperm, without which the embryo cannot develop. This is so-called pseudogamy.

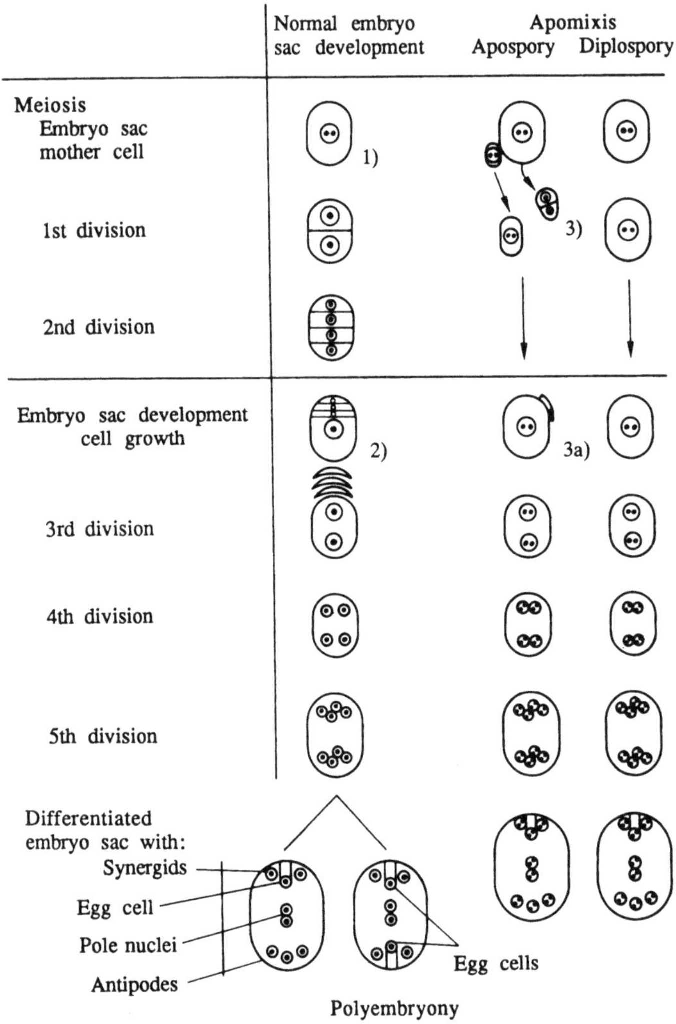

Fig. 37. Schematic development of the embryo sac. *1*) embryo sac mother cell; *2*) Macrospore; *3*) Degeneration of generative cells; *3a*) End of generative development

Another kind of apomictic multiplication is adventitious embryony, which is characterized by direct embryo development within a somatic cell, without the previous formation from the cell in question of an embryo sac with eight nuclei. If the somatic cell originated in the nucellus, the process is called *nucellar embryony*. For their further development, the adventitious embryos generally require endosperm from normal embryos resulting from amphimixis. Therefore, the seeds often contain several embryos (polyembryony), partly of apomictic and partly of amphimictic origin (Fig. 37).

Among apomicts, facultative apomicts, which partly form reduced and partly unreduced embryo sacs, and obligatory apomicts, exclusively forming unreduced embryo sacs deriving from apospory or diplospory, can be distinguished.

Apomicts of one species, which originated from apospory or diplospory, are almost always characterized by polyploidy, i.e., an increased number of chromosomes compared to the amphimicts of the same species. Species hybridization can also increase the number of chromosomes (for details, see Sects. 3.1.3 and 3.6.1). Since polyploid forms frequently show decreased fertility, above all due to disorders during gamete formation, apomictic seed formation provides advantages in selection, as long as this is already genetically fixed in the original material or evolves mutatively after polyploidization. The advantage of apomictic multiplication lies in the fact that extremely heterozygous and also aneuploid plants – forms that have some extra chromosomes besides the two pairing homologous chromosome sets – yield a constant progeny, as long as the apomixis is obligatory. In facultative apomixis, descendents can appear with differing genetic constitution due to amphimixis. In general, however, they can already be distinguished from the apomictic plants, which are genetically identical with the female parent, at an early juvenile stage, and can therefore be eliminated as undesired. The significance of apomixis for the maintenance of constancy in heterozygous forms has also been addressed in Section 2.3.

Recently, energy has also been devoted to breeding apomictic parent stock for apples. The advantage of seedling tree stocks over vegetatively propagated tree stocks is the small production cost; they are also free of virus diseases.

Apomictic tree stock is superior to normal seedling tree stock in uniformity, which is due to the genetic identity with the female parent, as in vegetative tree stock. Apomixis is not known in cultivated apples; however, several polyploid wild apple species multiply this way. The majority of wild apple species is incompatible with cultivated varieties when grafted. Therefore, the attempt must be made to select compatible apomictic tree stock deriving from crosses of wild species with cultivated varieties.

Nucellar embryony is especially widespread in citrus species and mango, and is already used in these cases to obtain virus-free tree stock and also for direct propagation of cultivated varieties. The embryonic nucellar tissue is particularly suitable as starting material for tissue cultures, with which rapid propagation can be achieved in vitro. Moreover, it is used as starting material in in vitro selection (Sect. 3.7.2). As already mentioned above, adventitious or nucellar embryony is characterized by polyembryony, i.e., several seedlings develop from one seed, among which are also amphimictic seeds from fertilization by foreign pollen, if these have not died earlier. In the juvenile stage, the hybrids must be separated from the seedlings, which are genetically identical to the female parent, as soon as they can be recognized as such.

Strong infestation with virus diseases and nematodes often causes trouble for breeders and propagators of strawberries. Therefore, the question arises as to whether apomixis, which sporadically occurs in this species too, could also be exploited here with economic success in breeding. The necessary combination of apomixis with economically valuable characters of choice varieties, however, will surely not be an easy task. So far, obligatory apomicts are not known in strawberry.

How can new varied starting material for the selection of new varieties be obtained in apomictic species? First, the breeder could rely on the occurrence of spontaneous mutations, as long as the apomictic multiplication is not simultaneously lost. In a selection among 360 individual plants of a stand of swamp meadow grass (*Poa palustris*) that were tested after cloning the individual root stocks, two exhibited a strongly different character in their development (H. Kuckuck, unpublished results). One of these had a widespread habitus and thus proved to be a negative variant. The other one, however, was characterized by higher leaf mass in the vegetative phase and remained constant in this positive feature when multiplied by seeds. Furthermore, this valuable mutant remained apomictic.

The expectation based on these findings, also passed on to other breeders, that by selecting twin plants new genetic variability for breeding new varieties can be created, has been confirmed: within a short period of time, new varieties were created in this simple fashion in *Poa pratensis* as well as in *Poa palustris*. The proportion of seeds from which twin plants derive amounts to 0.5–1%. Among these, there are 5–10% twin plants with one partner different from the starting material. New strains with improved characters can be selected by a performance trial. In most cases, they produce a uniform progeny caused by apomixis.

This variability most probably originated in somatic mutations and could thus be seen as parallel to the more numerous mutations in haploids and in vitro cell cultures (Sect. 3.1.4).

A breakdown of the apomictic systems can also be caused by radiation of seed with X-rays and, thereby, strongly varied starting material can be bred, in which new apomicts can again be successfully selected. Certainly many successes in improving various cultivated species by systematic application of apomixis in breeding can still be achieved.

3.4 Production and Use of Haploids

In Section 3.3, apomixis was described as an example for the asexual production of seeds. Such seeds arise either from diploid somatic (2n) cells that have displaced the reduced (1n), primary embryo sac cell (= apospory) or from embryo sac cells which are, due to some disturbance of meiosis, unreduced (= diplospory). In both cases, diploid seeds are formed without fertilization, a phenomenon that is called diploid parthenogenesis.

Regardless of the procedure used for haploid production, haploids themselves are of no practical use. Their genomes must double to result in doubled haploids. This happens spontaneously in the majority of cases during the in vitro culture steps. If the regeneration frequencies of 2n plants are relatively high, 1n regenerants are often just thrown away. If the total number of regenerants is, however, limited or if the proportion of 1n plants is very high, the haploids must be doubled. The most commonly used procedure is treatment with an aqueous solution of 0.5% colchicine, supplied via the root system to the entire plant or with cotton wool plugs that are placed after axillary buds are excised.

3.4.1 Parthenogenesis

Seeds can also arise from unfertilized (1n) cells. This latter process is called haploid parthenogenesis. Haploid parthenogenesis results in a haploid seed and, later on, in a haploid plant with a simple genome. The procedures that lead to haploid parthenogenesis are summarized in Fig. 37. Most often, haploid parthenogenesis results under in vivo conditions when, in addition to the reduced egg cell, other reduced cells of the embryo sac (e.g., one of the synergids) function as egg cells. This process is called polyembryony (Fig. 37). During the double fertilization of the two sperm cells with the egg cell and secondary embryo sac, such additional "egg cells" remain unfertilized. These cells may also be stimulated to form a seed, induced by the starting endosperm formation following the normal egg cell fertilization. From two egg cells, twins will issue, one of which is normally diploid, the other haploid (Fig. 37).

Haploid parthenogenesis may also be a result of abnormal pollen mitosis: if just one sperm nucleus is formed, this single nucleus can only fuse with either the egg cell or the embryo sac nucleus. If the egg cell remains unfertilized, it can form – induced by the normal endosperm development – a haploid parthenogenetic embryo.

A further irregularity of egg cell fertilization results in haploid embryos: after a normal, double fertilization the endosperm development is disturbed due to partial incompatibility of the parental genomes. This failure of endosperm development results in nonviable seeds in vivo. The abortion of the embryo may be prevented by excision and isolation of the ovule soon after fertilization, and then by culturing it in vitro on a suitable culture medium. The nonfunctional endosperm is replaced by a functional culture medium that allows further embryo growth. Depending on the degree of incompatibility, diploid hybrids are formed or one genome is partially or completely eliminated. If the male genome is completely eliminated, a haploid embryo is the result, a process similar to haploid parthenogenesis. In some plants, e.g., sugar beets, fertilization with irradiated pollen seems to be more effective for haploid induction than pollination with the pollen of wild species. Haploid parthenogenesis by chromosome elimination after wide crosses and subsequent male genome elimination is used, e.g., in the production of haploid barley plants. In this case, the cultivated barley (*Hordeum vulgare*) is fertilized with the wild barley species, *H. bulbosum* (Table 17).

Finally, an incompatibility of parental genomes resulting in haploid embryo formation is observed after crossing the same species but with different ploidy levels. In such pollinations, one of the sperm nuclei may develop in the ovule without fusion to the egg cell nucleus. Such a haploid embryo contains the

Table 17. General procedure for doubled haploid line production in Hordeum by interspecific cross and chromosome elimination (Kasha and Kao 1970)

Time Months	Days	Treatment	
1–4		Produce F_1 hybrid seeds	
4		Plant F_1 hybrids	Plant *Hordeum bulbosum* (2n)
6	0	F_1 *Hordeum vulgare* (2n)	*Hordeum bulbosum* (2n)
		Emasculation	Collect fresh pollen
	2	→ Pollination ←	
	3–5	GA_3 Treatments (1 drop/floret of 75 mg/l GA_3 for 2–3 days)	
	14–16	Dissection and culture of embryos in the dark	
	22–28	Culture of differentiated embryos at 12 h light circle	
8		Doubling of haploid seedling with colchicine (0.1% 5h)	
11		Harvest seeds from the haploids doubled	
11–14		Continue program for 3 months at rate of 100 haploids/week	
14		Propagation of seeds in a winter generation,	
20		Replicated field tests	
24		Quality check	
Total = 2 years		from parents to replicated field data on 600–800 double haploid lines	

cytoplasm of the female parent and thè nuclear genome of the male. This type of haploid formation already constitutes androgenesis and, because of its special location, is called ovule androgenesis.

3.4.2 Androgenesis

For applied purposes, androgenetic procedures are gaining in importance since, with increasing knowledge of this system, the artificial production of large

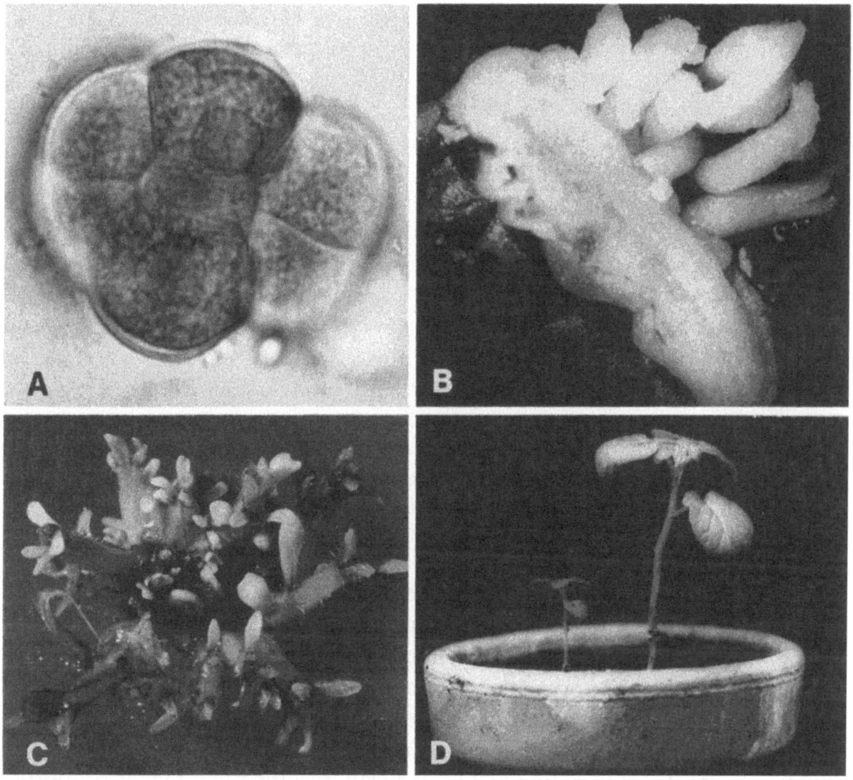

Fig. 38A-D. Development of haploid plants. ***A-C*** Androgenesis; ***D*** parthenogenesis. ***A*** Microspore after first somatic division; ***B*** embryoids shed from the anther; ***C*** young microspore plantlets still in contact with the anther; ***D*** twin plants of potato; the plant on the *left* was parthenogenetically produced and is haploid

numbers of haploids becomes economical. The fortuitous observation of Guha and Maheshwari (1964) that embryos grew from anthers of *Datura*, which were left over from a cytological experiment, established the basis for this important path in haploid induction. In this microspore androgenesis, the change from the generative to the sporophytic development is most often started by an equal, initial microspore mitosis (Figs. 38 and 39), which replaces the unequal division and leads to the formation of the vegetative and generative nucleus of the pollen grain. This equal division is a first step in the formation of a globular proembryo, which can develop into a normal embryo (since this embryo does not originate from a zygote, it should be called embryoid) or – and this happens quite often, probably under the influence of high phytohormone concentrations – into unorganized callus. Most commonly, the microspores are cultured within the entire anther, which is the reason why this method of microspore androgenesis for haploid production is quite often called anther culture. By carefully monitoring the preculture of the anther donor plants and by optimizing the culture media,

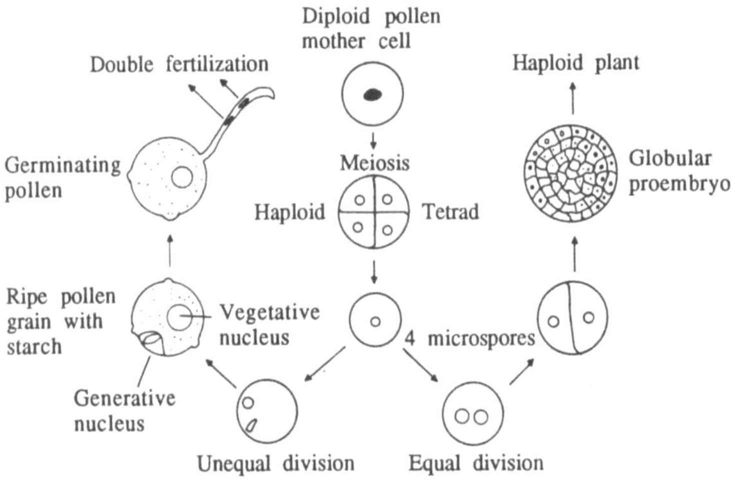

Fig. 39. Generative and somatic microspore development

in more and more cases an embryogenic pathway to haploids can be achieved. The haploid regeneration frequency in anther culture is also dependent upon the genotype; it is, however, more laborious to develop an optimal medium for a poorly regenerating genotype than it is to improve the in vitro reactions genetically. Nevertheless, one must avoid undesired genetic drifts. If the regeneration procedure avoids callus formation, the normally undesired genetic aberration due to gametoclonal variation will be reduced, resulting in a higher proportion of green lines with good agronomic fitness.

Increasingly, direct microspore culture replaces the anther culture procedure. The culture of microspores isolated from the anther is possible in crops such as potato, rapeseed, rice, barley, wheat, and maize. Isolated microspore culture apparently offers a more efficient system for regenerating random samples from microspore populations than does anther culture. Such a procedure, starting from single microspores passing through embryogenesis and developing at a high frequency into green plants, is most advantageous and desirable, because early selection in microspore populations then becomes possible. The genetic presence of desirable characters can be influenced by parent selection, and even complex reactions may be identified with simple selection systems, which are the equivalent of one step from the complex ones. Thus, this type of selection is superior to systems wherein the variability is only created by somaclonal variation (Sect. 3.1.4). Along this line, Ye et al. (1987) combined microspore culture and in vitro selection. They cultured microspores from a barley F_1 hybrid – a cross between a salt-tolerant and a normal, sensitive cultivar – in liquid media containing up to 0.8% Na_2SO_4. No progeny from the F_1 microspore cultures in the high salt medium was as susceptible as the susceptible parent. The results indicate that, in these experiments, sensitive microspores were eliminated. Lines exhibiting elevated levels of tolerance to salt derive from recombination rather than from gametoclonal variation. Microspores from rapeseed were also screened successfully for herbicide tolerance.

The use of isolated microspores offers the possibility of combining single-cell selection procedures (Sect. 3.7.2) with the advantages of a haploid system, which are in particular: absence of cross-feeding and chimerism, direct expression of all genes independent of their recessive or dominant nature, and the possibility of regenerating directly doubled haploid, homozygous lines. Furthermore, gene transfer to plant cells, which is increasingly becoming an important tool for applied breeding (Sect. 3.6.3), can make use of isolated microspore systems. The present lack of an appropriate vector system for gene transfer in monocots and the still limited success in regenerating cereal protoplasts make microspores a promising prospect for gene transfer and subsequent regeneration of transformed plants. The isolated microspore culture may allow, after microinjection or ballistic techniques, the detection of transgenic structures from which transgenic plants may be regenerated.

3.4.3 Haploid Steps in Breeding Programs

Today, doubled haploid lines are increasingly used in practical breeding programs, since haploids can be produced very economically. In barley, the regeneration frequency reaches 5–10%, based on the number of plated anthers, and similar successes are true for rapeseed. Since regeneration is no longer the central bottleneck in using haploids, other questions are assuming importance. One such question is: can haploids be used more economically in an F_1 generation or after some preselection in later generations? Depending on the nature of the starting material as well as on the breeding aims, one or more haploid steps may be applied during recombination. The classical F_n generation corresponds to the first haploid (A_1) generation, therefore selection in A_1 visualizes both the genotypic and the phenotypic characteristic. This allows the most commonly used partial bulk scheme to be substantially simplified (Fig. 40): the doubled haploid lines represent the final point of the seedling's bulk, which permits the immediate start of the pedigree part of the method. It has been demonstrated that haploids can also increase the breeding efficiency of the most complex breeding problem: the transfer of quantitatively inherited traits from distant genotypes to cultivars. For such complex breeding programs a procedure has been developed combining recurrent selection – which guarantees variability, and haploid selection – which guarantees reliable selection. This technique is called recurrent selection alternating with haploid steps. The degree of relationship of the two parents determines the number of backcrosses needed. If positive selection pressure is possible, the backcross can be done in the first generation on single plants without destroying this plant; if this is the case, then it will be possible to follow a breeding procedure such as is shown in Fig. 14.

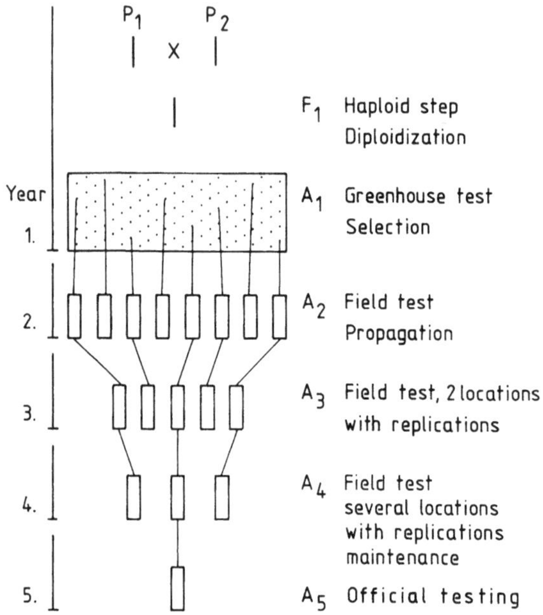

Fig. 40. Breeding scheme for the introduction of mono- and polygenically inherited traits from related varieties into inbreeders using a haploid step

3.4.3.1 Complex use of Haploids

Using potato as an example, a combined application of classical procedures and the use of haploids will demonstrate how in vitro procedures can increase the efficiency of combination breeding in vegetatively propagated crops.

The cultivated potato, *Solanum tuberosum*, is autotetraploid with 48 chromosomes (2n = 4x = 48). Breeding at the tetraploid ploidy level is difficult, particularly when a combination of quantitatively inherited characters is desired. As a first simplification, combination breeding is moved to the diploid (dihaploid) level (2n = 2x = 24). Such diploid forms are known from wild species and primitive cultivars. This first reduction of the ploidy level is done predominantly parthenogenetically, using the primitive cultivar *Solanum phureja* as pollinator. From successful pollinations, up to 40% haploid seeds can be detected. These seeds arise from parthenogenesis; nonhaploid seeds are mainly hybrids. The procedure can be further enhanced by marker genes, e.g., the anthocyanin color marker nodal band, already visible as the seed marker embryo spot, which in some *S. phureja* clones is combined with a high haploid-inducing capacity. Only seeds lacking such a spot may have a parthenogenetic origin and can be haploid. The selection in this combination is facilitated because, due to incompatibility between a 5x endosperm and a 3x embryo, no triploids are formed (triploid block). This block does not work, however, when in a subsequent combination diploid (dihaploid) *S. tuberosum* are pollinated with *S. phureja*. As a consequence, huge numbers of diploid hybrid seeds must be screened for the very small fraction of spotless monohaploids obtained (2n = x = 12). Because of this laborious selection, the androgenetic technique is much more efficient. Similar advances are the consequence of improvements in the androgenetic ploidy level reduction from 4x to 2x.

From monohaploid clones with genome A via doubled monoploids AA (dihaploids) and by doubling the dihaploids, tetraploids AAA can be regained in two steps. These tetraploids will be homozygous and will have the advantage that such potatoes may be treated as an inbreeder that can be propagated stably via true botanical seeds. Propagation via true potato seeds (TPS) is practiced in a number of countries, because TPS guarantee that most tuber-transmitted viruses will be absent. In areas where storage and distribution are of greater importance than homogeneity, TPS are in use and methodological problems such as transplanting and plant protection are solved. Most of the present TPS come from 4x × 2x crosses in breeding programs, where the potato is handled as an outbreeder yielding quite good homogeneity but no homozygosity. Homozygous TPS would probably not yield sufficiently, but they could be the basis for a hybrid breeding program. More probable is an approach to a synthetic tetraploid potato clone (Fig. 41): from four different tetraploid clones, superior in at least one trait, for example, clone I contains the resistance to potato virus X (PVX) on genome A_1; clone II, the resistance to potato virus Y (PVY) on genome A_2; clone III, a *Phytophthora* resistance (P) on genome A_3; and clone IV, a resistance to cyst nematodes (PCN) on genome A_4. The ploidy level of the four clones is reduced twice to reach the monohaploid level, and on each, selection is made for the presence of the respective resistance. After genome doubling, which normally

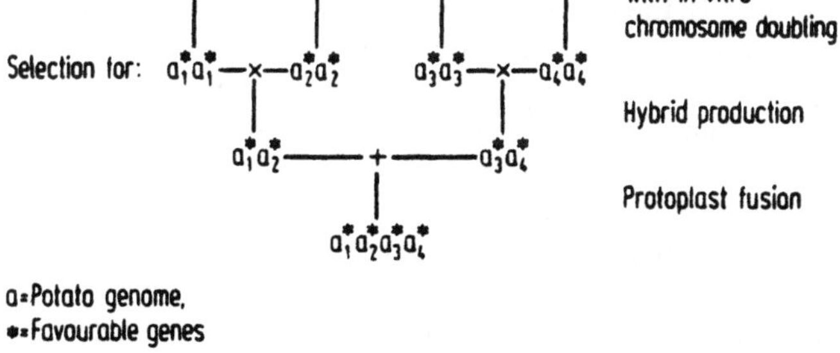

Fig. 41. Scheme for an analytical synthetic breeding procedure combining tissue culture techniques and conventional breeding steps in potato breeding

happens spontaneously, four homozygous doubled monohaploid (dihaploid) clones are available with one of the resistances. Such clones are classically combined by crossing yielding, dihaploid hybrids, which contain two of the resistances, e.g., $A_1A_1 \times A_2A_2$ gives A_1A_2 with PVX and PVY resistance, and $A_3A_3 \times A_4A_4$ gives A_3A_4 with *Phytophthora* and nematode resistance. Besides the combination of the resistances, such heterozygous hybrids quite often express strong heterosis. The final step is the somatic fusion of these hybrids to combine all four resistances and to reach, at the same time, the tetraploid ploidy level expression the highest amount of heterozygosity. Again, such hybrids express in a number of cases strong heterosis at the plant and tuber yield level, due to their increased level of heterozygosity. For details of the somatic fusion, see Section 3.6.2.

As long as the somatic fusion did not work, the return from the 2x to the 4x ploidy level was performed predominantly by meiotic doubling, a process that is reciprocal to haploid parthenogenesis. The 4x father is pollinated with 2x pollen, where sometimes just one diploid sperm nucleus is present. If this unreduced, single-sperm nucleus fuses with the normally reduced 2x egg cell of the tetraploid mother, a normal 4x zygote results. Depending on the process of origin of the unreduced (diplandroid) sperm cell, its genomes may be as heterozygous as the 2x clone, thus adding (as in somatic fusion) the whole male genome to the zygote. The process can also be done the other way around: 2x female and 4x pollinator. If the egg cell remains unreduced (digynoid) and is fertilized by a normally reduced sperm cell, again a tetraploid zygote is the result. This latter procedure is, however, less effective. In theory, an unreduced germ cell can come from both sides allowing, as in somatic fusion, genome addition and genome doubling.

The described analytical-synthetic breeding scheme is being investigated in potato breeding programs. First tetraploids with combined resistances are undergoing field trials.

3.5 Sex Inheritance and Its Consequences for Plant Breeding

Most species of flowering plants develop reproductive organs of both sexes on the same plant. These reproductive organs can be combined within one flower or can be distributed among separate pistillate and staminate flowers. In the first case, which applies to most cultivated plants, the species are called hermaphroditic and, in the second case, monoecious. Approximately 3% of flowering plants cannot be classified into either of these two groups. They develop separate sexual individuals, of which the male plants have only staminate and the female plants only pistillate flowers. These are the so-called dioecious species (for more information, see Frankel and Galun 1977).

In the dioecious species, such as hemp, hops, asparagus, and spinach, depending on the species, either the plants of the male, or female, have a higher economic value. Therefore, plant breeding attempts to develop species in which the economically less valuable sex types are either entirely eliminated or at least largely restricted in their numerical proportion. On the other hand, in monoecious species such as cucumber, pumpkin, maize, mulberry, melon, and squash, a shift of the numerical proportion in favor of the male or female flowers may be desired.

In order to solve these breeding tasks, a knowledge of the principles of sex inheritance and determination as well as the possibility of influencing sex expression by external factors are necessary. Since the sex type with better economic value depends on the corresponding plant species, and since the inheritance of the sex in dioecious plants as well as the flower types in monoecious plants are subject to certain variations, the breeding method to be applied is different. Not all dioecious and monoecious species can be discussed here, however. Therefore, only some of their typical representatives, such as cucumber, asparagus, and spinach, will subsequently be described in terms of the consequences in breeding resulting from sex inheritance.

Cucumber is originally a monoecious plant species. In East Asian varieties, purely female, and in the Australian variety White Lemon, andromonoecious plants, i.e., plants with staminate and perfect hermaphroditic flowers, have also been described. Thus, there are three different flower types in cucumber: pistillate, staminate, and perfect. According to the various distribution of these flower types on the plants, the different sex types result: monoecious, gynoecious, androecious, hermaphroditic, and andromonoecious. Hermaphroditic plants have only perfect flowers, gynoecious plants only pistillate, and androecious plants only staminate flowers.

Besides environmental factors and modifying genes, three major genes, Acr/acr, M/m, and A/a, are responsible for the inheritance of these different sex forms in cubumber (Table 18).

The monoecious plants are recessive at the Acr/acr locus. In general, they initially form staminate, then staminate and pistillate, and in the last phase, only pistillate flowers. The gene Acr promotes the female phase and, therefore, female sex expression by acceleration of the male and mixed stages. The dominance of Acr over acr is not always complete and is strongly dependent on genetic background and environmental factors. The development of hermaphroditic flowers is attributed to the recessive gene m. In the presence of acr in homozygous condition, analogous to monoecious plants, male flowers are also produced in addition to hermaphroditic flowers, so that andromonoecious plants are formed.

Table 18. Effect of sex genes in cucumber (Franken 1979)

	Acr A or a	acr A	acr a
M	Gynoecious (♀)	Monoecious (♂ and ♀)	Androecious (♂)
m	Hermaphroditic (⚥)	Andromonoecious (♂ and ⚥)	Androecious (♂)

M is completely dominant over m. Acr/acr and M/m are inherited independently from one another. Gene a is responsible for androecy. This gene is hypostatic to the gene Acr and has an effect in acr/acr plants only. Acr/acr and A/a are inherited independently from one another.

Statements concerning the effect of nongenetic factors on sex expression, such as temperature and light conditions, differ in the literature, which may be due to the different trial material used. Short days of less than 8 h and low temperatures and low light intensity promote formation of staminate flowers. In contrast, Lower and Edwards (1986) reported an increased number of staminate flowers due to high temperatures and/or long days. Under certain environmental conditions, particularly under stress conditions, gynoecious and hermaphroditic lines develop a number of staminate flowers, which differs from line to line.

The development of different flower types can be affected by the application of chemicals. For maintenance and multiplication of a gynoecious line for breeding and research purposes, treatment with silver nitrate is the cheapest and most effective method. Generally, treatment is carried out in the 1–4 true-leaf stage at concentrations between 100 and 300 ppm. In order to prevent leaf damage from occurring in some genotypes, silver thiosulfate $[Ag(S_2O_3)_2]$ can be used instead of silver nitrate, which is prepared immediately before use by adding excess (8:1) sodium thiosulfate to silver nitrate solutions. After treatment with silver nitrate, some of the lines also tend to form hermaphroditic flowers in addition to staminate flowers. (den Nijs and Visser 1980)

Fruits that have been produced by hermaphroditic flowers are sphere- to lemon-shaped and have no economic value. In addition to androecious plants, andromonoecious and hermaphroditic plants can thus also not be used for fruit production. Since in the remaining sex types only the pistillate flowers develop fruits, the yield potential of a plant is higher, the larger the proportion of pistillate flowers of this plant. Staminate flowers are not even necessary at all if femaleness is combined with the character *parthenocarpy*. Parthenocarpy is the ability to form fruits without fertilization. This character is inherited by an incomplete dominant gene (Pike and Peterson 1969) and can easily be introduced into the breeding material because of a relatively simple mode of inheritance.

With regard to breeding of varieties for mechanical harvest in pickling cucumbers, femaleness becomes even more important. In this form of harvest, the plants are harvested along with their vegetative parts, and the fruits are subsequently separated from other parts of the plant with the harvester. In contrast to picking by hand, which is done several times, mechanical harvest is a once-over harvest. Therefore, the plants must have developed a sufficient number of fruits in a concentrated form by the time of harvest. This is best guaranteed if the varieties consist of female plants only.

Thus, the aim in cucumber breeding is to develop mainly or purely female flowering varieties. This aim is reached in practice by the production of gynoecious F_1 hybrids, using a gynoecious line as seed parent. Except for androecious plants, which are difficult to multiply, and thus disregarding the effect of gene a, four of the above-mentioned sex types of cucumber remain, having the following homozygous genetic composition:

- gynoecious: MM AcrAcr
- hermaphroditic: mm AcrAcr
- monoecious: MM acracr
- andromonoecious: mm acracr

All four of these sex types can theoretically serve as male parent in the production of a gynoecious F_1 hybrid. The induction of staminate flowers in the multiplication of the gynoecious seed parent, as well as in the crosses "gynoecious × gynoecious," is done by $AgNO_3$ treatment. The F_1 hybrids produced by the crosses "gynoecious × monoecious" and "gynoecious × andromonoecious" (MM Acracr and Mm Acracr, respectively) can display a varying amount of staminate flowers due to the partial dominance of Acr over acr, also depending on environmental conditions (Tatlioglu 1983). This is troublesome as far as the combination with parthenocarpy is concerned or, in non-parthenocarpic varieties, as far as the calculation of the seed portion of pollinating plants to be added is concerned. The fruits formed in genetic parthenocarpic varieties after fertilization become misshapen, are of no economical value, and lead to loss of profits.

The F_1 hybrids produced by the crosses "gynoecious × gynoecious" and "gynoecious × hermaphroditic" (MM AcrAcr and Mm AcrAcr, respectively), on the other hand, have a considerably more stable female sex expression. There are indications that heterozygosity of the M/m locus in Mm AcrAcr hybrids leads to a more stable femaleness of the plants in comparison to the gynoecious lines used in the production of such hybrids. Since the gene m is completely recessive, the negative influence of this gene on fruit shape is completely compensated for in Mm plants. Thus, the hermaphroditic sex type has a special significance as male hybrid parent. With its usage the unreliable aspects of chemical induction of male flowers on female plants, e.g., in seed production in the field, can be avoided.

In the dioecious species, such as asparagus, spinach, and hemp, female and male plants are formed at a ratio of 1:1. The presence of two, depending on the plant species, more or less structurally differing chromosomes, the so-called sex chromosomes (termed X and Y), are responsible for maintenance of this ratio. One of the sexes is homogametically XX and the other sex, heterogametically XY. The homogametic sex forms one type of gametes, namely, those only containing X, whereas the heterogametic sex forms 50% gametes with X and 50% gametes with Y. Thus, the cross female XX × male XY results in 50% XX (female) and 50% XY (male) again.

The sex chromosome pair contains a sex-deciding factor – a gene or a chromosome segment – that suppresses formation of the sex organs of one of the sexes, so that only flowers with organs of the opposite sex are produced. In most cases, heteromorphy between the X and Y chromosomes is so strong that, in meiosis, a tighter pairing and thus an exchange of segments containing sex-deciding genes, by crossing-over, can largely be excluded. Therefore, the effect of the sex chromosomes is also guaranteed for subsequent generations.

Furthermore, there are so-called sex-promoting genes in the autosomes – which are the chromosomes remaining after exclusion of the sex chromosomes from the entire chromosome set – of every individual, responsible for the formation of male as well as female sex organs, so that in every individual of a dioecious species a bisexual potential is basically present. Like trigger mechanism, the XY mechanism decides whether the male or the female organs are promoted or suppressed in the development of an individual. If, however, the effect of sex-deciding genes, which are not only located on the sex chromosomes but can also be on the autosomes, is altered by mutation or recombination, the XY mechanism may lose its effect as trigger for the formation of one sex and suppression of the opposite sex. If this is the case, then sexual intermediaries or monoecious forms can occur. Crosses between more distantly related races with different genetic composition with respect to the sex genes may also lead to the loss of balance of the gene systems and, consequently, result in the formation of monoecious plants.

In asparagus, too, sex inheritance is regulated by the XY mechanism. The female plants are homogametic XX, the male plants heterogametic XY. The male plants have a higher yield than the female individuals. They live longer and yield earlier. Accordingly, plant breeders make an effort to develop purely male asparagus varieties.

Occasionally, a third sex type, andromonoecism, occurs in asparagus, which is of the sex chromosome constitution XY just like the male plants, but is able to produce perfect flowers and seeds besides male flowers. The andromonoecious plants can be selfed or crossed with one another. In the progenies, a segregation ratio of 1 XX (female) : 2 XY (male) : 1 YY (male) is to be expected. All male plants with the new combination YY, which are also called super males, can be used as pollinators of normal XX females. As a consequence, the resulting progeny must exclusively consist of XY males. Since the homogametic YY male plants cannot be distinguished from the XY male plants in a phenotypic respect, their selection from the segregating progenies is difficult. Trisome analyses indicated that the asparagus chromosome L5 is the sex chromosome, however, the X and Y chromosomes cannot be definitely differentiated, so that a cytological identification of XY male plants is not possible. They must be selected by test crosses with female plants. Since asparagus plants flower about 3 years after sowing, the results of these testcrosses can only be obtained after a corresponding period of time.

In order to produce seed on a larger scale for a variety consisting of male plants only, the identified YY male plants must be multiplied, which is of necessity only possible by vegetative propagation. Asparagus plants, however, are not easily cloned in vivo. On the other hand, this is possible in vitro, although not reliable enough for all genotypes. The methods of in vitro propagation and maintenance of asparagus, also with regard to breeding of purely male varieties, have been described in detail by Ellison (1986).

As described in Section 3.4.2, completely homozygous genotypes can be produced with anther culture. This also applies for the sex chromosomes. After reduction division within the anther, the X or Y chromosomes are present at a ratio of 1:1. Both types of microspores can be stimulated to develop androgenetically into haploids. After duplicating the chromosome set, homozygous XX female plants or homozygous YY super male plants are obtained from the haploids. These super male individuals can now be multiplied and maintained in vitro by tissue culture. With this method, no XY male plants are formed and tedious testcrosses can be omitted. The two homozygous groups XX and YY that occur can phenotypically be distinguished by the flower.

The character andromonoecism is also determined by genetic factors and can thus be transferred to the YY super males and their progenies from the crosses XX × YY. With this, the economic value of purely male varieties is diminished, however, because the perfect flowers of the XY males form seed just like the flowers of female plants, and therefore require important nutrients for their production, which are probably lost in the formation of the shoots. Breeding of purely male varieties without the tendency to form berries should be possible. For this, andromonoecious forms with a low andromonoecious value should be used and super males and female cross partners should be selected, whose autosomes have genes preventing the development of andromonoecism in the F_1.

Spinach is generally classified as a dioecious species. The maintenance of dioecious populations is realized by the XY mechanism. The female plants are homogametic XX and the male heterogametic XY. As could be definitely shown in trisomic analyses, the longest chromosome of the genome is the sex chromosome (Löptien 1979a).

The flower shoot of the female plant is more or less compact and is covered with leaves to the top. The female flowers have no peduncles and are clustered in the axils of the leaves. The inflorescence of the male plants forms a loose, false ear without leaves carrying the flowers on peduncles in clusters. The male plants are undesired for cultivation, because they start bolting up to 10 days earlier than the female plants, and thus determine the time of harvest. Consequently, there is a low yield, since at that point in time the female plants have not yet reached their full leaf yield. As soon as flowering starts, the content of oxalic acid in the male plants increases, thus restricting further use of the plants as a vegetable. Male plants also exist, which have female habitus and do not display the disadvantages of the normal male plants.

Depending on the varieties or origins of populations, monoecious forms may also occur in varying frequencies, in addition to male and female plants. The monoecious forms have a female habitus. Their proportion of male flowers can strongly vary.

To avoid the disadvantages of the male plants with male habitus, plant breeding aspires to develop varieties without any male plants, i.e., consisting only of female plants and monoecious forms. In such varieties, male plants with female habitus can also be tolerated.

Starting from a few monoecious plants having dioecious origin, it is possible to obtain populations that are virtually free of male plants, by continuous mass selection. Starting with a dioecious variety with 1.5% monoecious forms only, Zoschke (1956) succeeded in developing a population consisting of 88.4% monoecious plants, 10.5% female plants, 0.7% male plants with female habitus, and only 0.4% male plants with male habitus within six generations. Such a population represents the original material for the breeding of a monoecious variety. It can be improved in yield and quality characters by use of known breeding methods.

The result obtained by Zoschke, that a dioecious variety can be converted to a monoecious one by continuous mass selection, can be explained by the relatively strong effect of the sex genes in the autosomes on sex expression in spinach and/or the relative sensitivity of the balance of the XY mechanism. Besides the classical dioecicism, XX for the female and XY for the male sex, a second possibility can arise from recombination of the sex-deciding genes: XY constitution for the female and YY constitution for the male sex (Dressler 1973).

Another general intention in spinach breeding is the production of hybrid varieties. In addition to the known advantages of hybrid varieties, it is also simpler to combine monogenic, dominant resistance genes against the three races of *Peronospora spinaciae* and cucumber mosaic virus in the F_1 hybrid. The main problem with hybrid breeding is the development of a purely female seed parent.

One method of hybrid seed production is the application of a dioecious line as female line, from which the male plants are rogued. The procedure, especially applied in the USA and Europe, is initiated from two dioecious lines with good specific combining ability, which are cultivated in alternating rows (6-8 rows of female line and 1–2 rows of pollinator line each). At the beginning of flowering, the male plants of the female line are hoed, so that only pure hybrid seed is yielded later on. Thereafter, a dioecious hybrid variety is developed. If a corresponding monoecious pollinator with sufficient pollen production is used, a monoecious hybrid without the already mentioned disadvantages of the male plants can be produced. After such crosses, it is also possible to produce a dioecious hybrid with male plants of female habitus by application of the dominant gene L postulated by Kubicki (1982) for abundant foliage.

The roguing of male plants when using a dioecious line as seed parent must occur several times, because flowering does not start uniformly and thus increases production costs. Since this cannot always be carried out with sufficient certainty, using a pure female line as seed parent has been considered, thus omitting the roguing of male plants. Long days and high temperatures promote formation of male flowers in spinach. Accordingly, monoecious plants tending to form few male flowers are selected in practical breeding under environmental conditions that promote maleness. The progenies deriving from such monoecious plants form female flowers in cooler climates and serve as seed parents in hybrid seed production. They can be pollinated by another monoecious line with sufficient pollen production or by a dioecious line that has good specific combining ability. Since spinach has a relatively high seed multiplication (sowing: 6 kg/ha, seed yield up to 2000 kg/ha), the propagation of the female line is also possible in the greenhouse.

Although male sterility in spinach exists, it has not been possible thus far to produce pure female lines consisting of male-sterile monoecious line (= female function of the individual plants) or of male-sterile males and normal females in dioecious lines (= female function of the line), because this kind of male sterility is controlled by nuclear genes only and no cms system is available (Kobabe 1972).

3.6 Other Forms of Genome and Gene Combinations

The use of interspecific crosses in breeding is often associated with many difficulties, e.g., in the production of hybrids, and because of the diverging cytogenetic behavior of the following generation in comparison to the more regularly segregating progenies of hybrids between varieties. These difficulties in interspecific crosses are often of different degree, and thus the possibilities of overcoming and successfully using them in breeding are to be differently evaluated. Among others, they are based on the fact that the term "species" as an independent, systematic unit cannot always be definitely determined. While formerly, species were mainly described and distinguished from one another by systematists on the basis of morphological and anatomical investigations, recently evaluation characteristics from other disciplines such as cytology, ecology, and molecular biology have also been applied. Thus, establishing the chromosome number of a species and determining their homologies, i.e., their structural equivalents, expressed in the formation of bivalents during meiosis, play an important role in the evaluation of systematic units. Adaptation to special ecological conditions is also of great significance, which in turn promotes the formation of isolated reproductive populations. The development of such separate reproductive populations is increased by the effectiveness of their different isolation mechanisms, such as flowering periods that strongly differ in time, cleistogamy, cross-sterility, and others. Thus, a multiplicity of characters is crucial for systematic assessment. Whether a group of plants, differing from an existing species in some of the characters just mentioned, is described as an independent new species or is regarded as a subspecies or variety of an existing species often depends on the researcher. The proper task of the systematist is first of all to classify clearly natural diversity of forms. This is in any case of great use, even if methodical perfection cannot be reached in the systematic respect. Since the variation of organisms is in the final analysis quantitative rather than qualitative in nature, transitional series are present in nature, and thus drawing borderlines between systematic units is often particularly difficult. As shown in Fig. 57, today comparisons of molecular markers and polymorphisms on the DNA level increasingly offer an additional tool to prove the systematics right or wrong. How to use this technique is described in Section 3.7.3.

It is thus not surprising if the breeder sometimes discovers that interspecific crosses succeed without difficulty and that the F_1 hybrids are fully fertile, but the parental characters of the following F_2, especially in their quantitative expression, segregate in such a complicated manner that segregation analyses with the help of the classical Mendelian method by formation of clearly definable classes of characters, are not possible. To achieve the desired combination success in such complicated segregations, the backcross method must be applied. After each backcross a selection on the characters of the "donor parent" (= non-backcross parent) must take place in the subsequent generation. Such a situation exists, e.g., in the transfer of *Cladosporium* resistance of the small-fruited wild tomato, *Lycopersicum pimpinellifolium*, to the large-fruited, but *Cladosporium*-susceptible cultivated tomato, *Lycopersicum esculentum*, as has already been described in Section 2.2.2.1. If, on the other hand, a quantitative character with polygenic inheritance from each of both parents is to be combined, application of convergence breeding, a modified form of the backcross method, which has already been dealt with in the above-mentioned section, is recommended. Actually, in the cited example of an interspecific cross, breeding processes are involved that do not basically differ from those of a normal variety cross of one and the same species. But obviously complicated segregations of polygenically determined quantitative characters make breeding success much more difficult.

In general, however, manifold difficulties must be overcome in interspecific crosses and their use in breeding. How this can be done will be illustrated further on with some examples. These classical procedures are opposed to only recently developed methods, which make use of special techniques to directly fuse protoplasts of two combination partners. The combination of complete genomes is further improved by asymmetric fusion and cybrid production (Sect. 3.6.2), and is even further perfected by the possibilities offered by gene transfer techniques (Sect. 3.6.3).

3.6.1 Sexual Interspecific and Intergeneric Hybrids

If the species to be crosses are characterized by different requirements of temperature, atmospheric humidity, or photoperiod, their flowering times will often be so widely divergent that already for this reason the crosses cannot be carried out. A convergence of flowering periods can be attempted by early sprouting of those species with longer vegetative times or by cutting back early flowering species. If a winter type of a cereal species is crossed with a summer type, both will be sown in spring. In addition, however, the seeds of the winter type to be crossed must be vernalized, i.e., exposed to the low temperatures needed for the transition from the vegetative to the generative phase (see Sect. 4.2). If, on the other hand, the species to be crossed differ in photoperiod, e.g., the cultivated potato and its related primitive types and wild species, then the wild species to be crossed from regions with short days must, in the northern degrees of latitude with long-day conditions, be darkened in such a way that it is only exposed to light for 8–12 h, as required. Only by this treatment will both species flower at the same time.

The success of a cross between two different species is also strongly dependent on the cultivars (genotypes) used. Thus, crosses of the wheat variety "Chinese Spring" with different inbred lines of rye resulted in uniformly high average seed set of 82%; after crosses with the wheat variety "Kolibri", however, the seed set of the same I lines varied, depending on the inbred line, between 0 and 50%. Therefore, a larger number of different genotypes of both species must be used in interspecific crosses, before a final decision can be made as to whether a cross between two species is practicable or not. Furthermore, it is important that the crosses are carried out reciprocally. Thus, the reason for the failure of a cross can, e.g., lie in the fact that the species used as male parent is short-styled. Therefore, the pollen tube may not be in a position to penetrate the style and reach the embryo sac of the long-styled species used as female parent.

The physiological interactions between pollen tube and style tissue are thus interrupted in this case. These disorders are particularly expressed if polyploid species with different chromosome numbers are used. For example, in *Datura* crosses the species with more chromosomes has the higher seed set if used as female parent; in *Triticum* and *Avena* it is just the opposite. The growth of foreign pollen and thus cross-pollination can also be facilitated in that its own pollen is added, which makes it easier, so to speak, for the foreign pollen to grow through the style and reach the embryo sac. Therefore, there will be plants deriving from the species used as female parent in the progeny of the majority of the plants, i.e., from self-pollination, but occasionally also a few hybrids will appear due to cross-pollination.

As far as seed formation in an interspecific cross is concerned, its development is influenced by the mutual ratio of the chromosome number within the embryo, the endosperm, and in the tissue of the ovary surrounding it. Normally, this ratio is 2x : 3x : 2x, i.e., embryo and ovary are diploid, whereas the endosperm is triploid due to the fertilization of the two pole nuclei of the embryo sac by the generative nucleus of the pollen. This mutual ratio is disturbed in crosses between species with different degrees of polyploidy, i.e., it is different in reciprocal crosses, as can be seen in Table 19.

As a consequence of the altered mutual ratios between the chromosome numbers of embryo, endosperm, and ovary (female tissue), morphological and physiological changes in the seed formation can occur, which have an effect on the development of the seed and its germination ability and are, namely, different in reciprocal crosses. However, there is no fixed rule as to whether use of the species with more or less chromosomes should be preferred as female parent. The disorders in development first occur in the endosperm. The breakdown of endosperm development is also one of the causes for the cessation of embryo development after regular fertilization and apparently normal initial development and finally the degeneration of the embryo. Therefore, the embryo must be dissected in time and further cultivated in vitro.

Table 19. Chromosome number of embryo, endosperm, and ovary after crosses between different ploidy levels (After Kappert and Rudorf 1958)

♀ × ♂	Embryo	: Endosperm	Ovary	Embryo / Endosperm	Embryo / Ovary	Endosperm / Ovary
2x × 2x	2x	: 3x	: 2x	0.67	1.00	1.50
2x × 4x	3x	: 4x	: 2x	0.75	1.50	2.00
4x × 2x	3x	: 5x	: 4x	0.60	0.75	1.25
2x × 6x	4x	: 5x	: 2x	0.80	2.00	2.50
6x × 2x	4x	: 7x	: 6x	0.57	0.67	1.17

The F_1 hybrids will be uniform, as long as the parent species used are homozygous. In their development, the hybrids can display all transitions from luxuriance to normal development to sublethal dwarf types. In the expression of their traits they will, in general, be in an intermediate position between the parent species or be completely identical to one of the parents. Pseudohybrids, which are identical to the female parent, are formed if apomictic species are used. However, they can also occur in amphimicts such as *Nicotiana*, *Solanum*, *Fragaria*, if a parthenogenetic development of the egg cell is induced by the foreign pollen. If the parthenogenetic development can traced back to unreduced egg cells, then all progenies are identical with the female parent. In haploid parthenogenesis, diploid plants can also occur in addition to haploid ones, which are formed from the doubling of the chromosome number of the egg cell; they are therefore completely homozygous, but need not be genetically identical to the female parent. A so-called prickle-pollination by foreign pollen could therefore be a very elegant method for the creation of homozygous inbred lines. Pseudohybrids identical with the male parent are also known; their development has not yet been fully explained. However, either the gametophyte without egg nucleus is able to further develop itself after penetration of the generative pollen nucleus, or it is also conceivable that after fertilization, the female genome is again eliminated from the zygote.

The further breeding work on interspecific hybrids is largely dependent on their degree of fertility. If they are fully sterile in both sexes, they can only be used as amphidiploids after doubling of the chromosome number (see Sect. 3.1.3). This is a rare case, however; at the least the female sex has a certain fertility, so that backcrosses of the hybrid with both parent species are possible.

During meiosis, a wheat-rye hybrid with 28 chromosomes (wheat $2n = 41$ with the genome AABBDD, rye $2n = 14$ with the genome RR) seldom forms gametes containing either all 21 wheat chromosomes or all 7 rye chromosomes along with a varying number of chromosomes of the other species, so that only little seed set can be expected in backcrosses. Plants with the two complete genomes of the original species seldom occur in the following generation. Therefore, it is better to use the amphidiploid hybrid $2n = 56$ chromosomes (genome formula AABBDDRR) for the backcrosses, since it only forms gametes with the complete wheat and rye genome, so that either 21 wheat bivalents or 7 rye bivalents are formed in the progeny of backcrosses. The numerical portion of the univalents varies and they are completely eliminated by repeated backcrosses. Thus, the two original species wheat or rye can be formed again in their genomic constitution, with the difference only that the rye genomes are now embedded in a foreign plasm, i.e., a wheat plasm. This new interaction between the hereditary elements of the rye nucleus and the plasmatic hereditary elements of the wheat plasm leads to a modified formation of characters. In the so-called "alloplasmic" rye, the anthers are occasionally more or less deformed and modified into carpel-like structures and the lodiculae into glume-like structures, so that the flowers are now cleistogamous. Due to the influence of the foreign plasmons on the rye genome, an inhibitory effect on the vegetative development can be observed, which occurs in addition to the inbreeding depressions after selfing of such backcross hybrids. As described in Section 2.3.2, the introduction of a nucleus into a foreign plasmon can also lead to male-sterile types, which are used as female cross parents for the development of hybrid varieties.

The "passage" of one species through an interspecific or intergeneric hybrid with the help of several backcrosses, which has just been described, can thus be of significance for the breeder. However, it is not possible to transfer single specific characters or genes from one species to another with this method, if the chromosomes between both species fail to form pairs and do not display crossing-over due to lacking homology. In this case, so-called *addition* or *substitution hybrids* have to be produced, i.e., a further chromosome pair of the donor species is added to the species to be improved or a chromosome pair of the species to be improved is replaced by one of the donor species. How is such an addition or exchange of a chromosome pair experimentally carried out? If, for example, an amphidiploid wheat-rye hybrid (2n = 56) is backcrossed with a wheat (2n = 42), the backcross product with 2n = 49 chromosomes forms 21 bivalents (= 21^{II}) and 7 univalents (= 7^{I}) during meiosis, which are distributed at random among the gametes containing a complete wheat genome, insofar as they are not lost in the process of gamete formation (Fig. 42). After a further backcross or selfing, plants with 43 chromosomes are sought in the progeny which, after another selfing, also result in plants with 44 chromosomes, i.e., wheat containing an additional

Amphidiploid hybrid between wheat and rye		Wheat
Genome AA BB DD RR		Genome AA BB DD
2n = 56	×	2n = 42
n = 28		n = 21

RK = AA BB DD R
2n = 49
Meiosis $21^{II} + 7^{I}$

Gametes: n = 21 + 0.21 + 1.21 + 2.21 + 3.21 + 4.21 + 5.21 + 6.21 + 7
Zygotes formed, among others, by the gametes:
n = (21 + 1) + (21 + 0) = 2n = 42 + 1

Wheat with one rye chromosome: 2n = 43
Gametes: 21 + 0.21 + 1
Zygotes formed by n = (21+1) + (21 + 1) = 2n = 44
↓
fixed addition hybrid

Fig. 42. Formation of an addition hybrid in wheat

chromosome pair from rye. Since rye has n = 7 chromosomes, seven different addition hybrids can be produced (Jenkins 1963).

For the production of substitution hybrids, so-called monosomic wheat is used as female cross parent, in which a chromosome pair should be substituted. The hexaploid wheat *T. aestivum* (2n = 42) is of an allopolyploid nature and contains the genomes A, B, and D with seven chromosomes each. The seven chromosomes of genome A are designated 1A to 7A, those of genome B, 1B to 7B, and those of genome D, 1D to 7D. One chromosome of each genome forms a group of homoeologous chromosomes, which are chromosomes that are so much alike that they can partially replace one another in function, e.g., 1A, 1B, 1D. Now if chromosome 1A is eliminated, the plants are changed but remain vital, since chromosome 1B or 1D can partially take over the function of 1A. If *one* chromosome is lost, a monosomic plant is formed and, if a chromosome pair is lost, a nullisomic plant. Twenty-one monosomic or nullisomic wheat types can be produced in wheat with 42 chromosomes.

To establish a substitution hybrid, a monosomic line ($20^{II} + 1^{I}$) is crossed with an addition hybrid ($21^{II} + 1^{II}$), the formation of which has already been described above. This results in F_1 hybrids, which are monosomic for one wheat and rye chromosome each (Fig. 43). These are in turn backcrossed with the addition line as male parent. This backcross progeny must be cytologically investigated on plants, which are expected to be disomic with regard to the rye chromosome and monosomic with regard to the wheat chromosome, if a monosomic chromosome has not been lost during the formation of the gametes. For the self-fertilized progenies of these monosomic-disomic plants, types must also be obtained that

Monosomic wheat line
2n = 41

Meiosis: $21^{II} + 1^{I}$
Gametes: n = 20 + 0.20 + 1

×

Wheat addition hybrid
2n = 42 + 2 rye chromosomes

Meiosis: $21^{II} + 1^{II}$
Gametes: 21 + 1

2n = 42

Meiosis: $20^{II} + 1^{I}$ (wheat) + 1^{I} (rye)

Gametes: n = 20
n = 20 + 1 (W)
n = 20 + 1 (R)
n = 20 + 1 (W) + 1 (R)

Zygotes formed by: (20 + 1 R) + (20 + 1 R) = 2n = 42

↓

Substitution hybrid
2n = 42 (40 wheat and 2 rye chromosomes)

Fig. 43. Formation of a substitution hybrid in wheat

are nullisomic as to the wheat chromosome but disomic as to the substituted rye chromosome. Consequently, the substitution hybrid thus obtained has 2n = 42 chromosomes like normal wheat and forms 21 bivalents.

The individual addition and substitution hybrids are of varying stability. Their value for the breeder is furthermore dependent on whether just *one valuable* character, e.g., resistance to parasitic fungus diseases, is transferred along with the added chromosome of the other species or whether disadvantageous consequences in the developmental process of the hybrid are also due to the foreign chromosome.

Disadvantages of this kind can largely be compensated for by an ingenious method that was successfully introduced into cytogenetic practice by Sears (1956): the translocation method. With this method, an attempt is made to split off a piece of the chromosome of the donor species containing the corresponding gene for the trait of interest and translocate it to a chromosome of the species to be improved.

Sears succeeded in doing this for the first time in the transfer of resistance to brown rust from *Aegilops umbellulata* to *Triticum aestivum*. The monosomic or disomic addition hybrids (2n = 42 + 1 or 2n = 42 + 2) are resistant to brown rust, but they have other disadvantageous characters, so that they cannot be used in cultivation. Among the addition hybrids, there was also a type containing the *umbellulata* chromosome as so-called isochromosome. An isochromosome is formed if one arm of the chromosome breaks off directly at the centromere and is lost, so that initially a telocentric chromosome (with a terminal centromere) is developed. If by chance the two chromatids of the telocentric chromosomes are not separated from one another to migrate to two daughter cells, but one chromatid performs a rotation of 180° around the centromere, a new chromosome with two arms that are mirror images of one another is formed with a median centromere in the middle, thus having the gene for resistance to brown rust at two loci (Fig. 44).

If, during an X-ray treatment of pollen cells containing such isochromosomes, single chromatin fragments are broken off through the irradiation, the chance that fragments *containing* the resistance gene will break off and attach themselves to the wheat chromosome or will be introduced into the wheat chromosome after previous breakage of the same is twice as high as in the normal monosomic types. Sears actually succeeded, after irradiation of male gametes

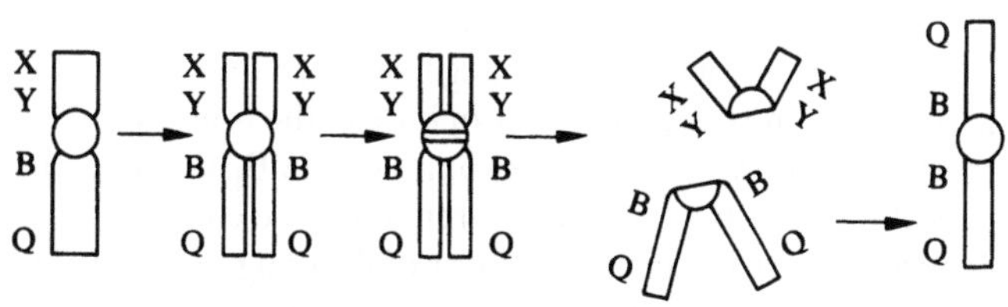

Fig. 44. Formation of an isochromosome

with an isochromosome, in inducing a wheat plant with 42 chromosomes that forms 21 bivalents and is resistant to brown rust. It has no disadvantageous characters, so that it could be introduced as a variety. Thus, a chromosome fragment containing the resistance gene has broken off due to two breakages of the *umbellulata* isochromosome, and this has been intercalarly introduced into a wheat chromosome after a further breakage of the latter. The introduced chromosome fragment is so small that the bivalent formation is not disturbed and also no other genetic disadvantages occur.

If crosses between two species or genera are not directly successful as is, e.g., also the case in the present cross between *Triticum aestivum* and *Aegilops umbellulata*, cross barriers can be overcome by building a so-called bridge. Therefore, Sears crossed the tetraploid wild wheat *Triticum dicoccoides* (2n = 28 with the genomes AABB) with *Aegilops umbellulata* (2n = 14 with the genome C^UC^U and initially obtained the hybrid ABC^U with 21 chromosomes. The amphidiploid $AABBC^UC^U$ with 2n = 42 chromosomes obtained from this can now be crossed with the hexaploid wheat. The cross $AABBC^UC^U \times AABBDD$ likewise results in a hybrid with 42 chromosomes of the constitution $AABBC^UD$, forming 14 bivalents and 14 univalents, and has a relatively large number of fertilizable egg cells. By backcrosses with *Triticum aestivum* AABBDD, plants with 42 chromosomes and resistance to brown rust can be simultaneously selected, which additionally contain a C^U chromosome as a monosome or disome. Such plants formed the source in the above considerations on the translocation of a chromatin fragment from the addition hybrid.

Interesting genetic breeding situations occur if species with different degrees of polyploidy are crossed with each other, i.e., differing in one genome but having their remaining genomes in common. This is, e.g., the case if the diploid *Tagetes erecta* (2n = 24) is crossed with the allotetraploid *Tagetes patula* (2n = 48). The latter is an amphidiploid species, which has likely been formed by the cross between *Tagetes erecta* (genome AA) and the diploid *Tagetes tenuifolia* with subsequent chromosome doubling, and consequently contains the genomes A A B_p B_p. *Tagetes patula* and *Tagetes erecta* have the A genome in common; it is structurally the same in both species, so that pairings between the A chromosomes occur within the hybrid *patula* × *erecta*. Since they differ in many alleles of their genes, they have been designated A_p and A_e genomes (Fig. 45). As can be seen from Fig. 46, an even distribution of the partners of the homologous A chromosomes on the two opposite poles takes place during metaphase, whereby a recombination of the chromosome sets occurs, in which the A_p and A_e chromosomes no longer stay together in one chromosome set each, but rather a mutual exchange takes place. It may also be assumed that by crossing-over, a fragment exchange between one chromatid each of the A_p and A_e chromosome occurs in the prophase in the four-strand stage. The same figure further shows how the chromosomes of the B_p genome are distributed to both poles at random as univalents, or remain at the equatorial plate and are lost again because they are not included in one of the daughter cells. Consequently, gametes are developed containing 12-24 chromosomes, depending on the number of univalents of the B genome, whereas the A genome with 12 chromosomes is

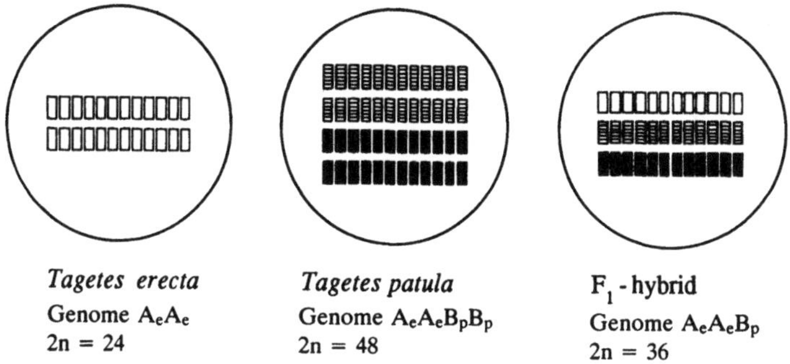

Fig. 45. Genome constitution of *Tagetes erecta* and *T. patula* and the F_1 hybrid (Eresen 1966)

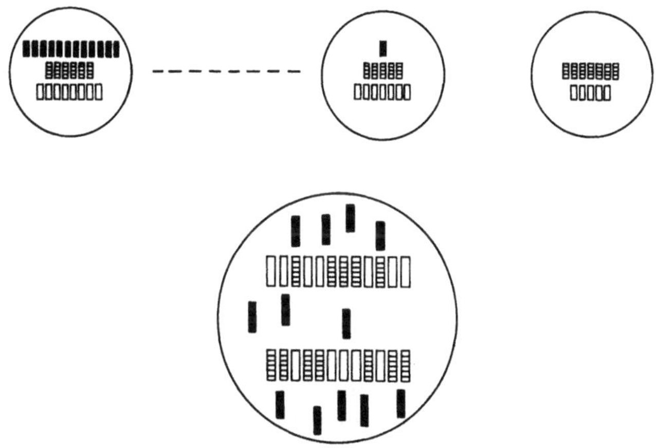

Fig. 46. Meiosis of the F_1 hybrid and formation of gametes with A_e chromosomes and with a chromosome number of the B genome varying from 0 to 12 (Eresen 1966)

present in full strength in all gametes. If it is assumed that all gametes formed in this way are developed with a varying chromosome number in both sexes at the same frequency, and further that the same chances exist for the formation of a zygote, plants with a chromosome number of 24–48 chromosomes in the F_2 generation of the triploid hybrid are to be expected at the frequency shown in Fig. 47. The actual chromosome numbers strongly deviate from the expected ones, however. The aneuploid gametes have a different viability. The survival ability is greater with only a few additional chromosomes, as is likewise the case with a more or less complete B chromosome set. As a result, a downward or upward adjustment to the euploid original species occurs in the following generations; that is, the pure parent species *Tagetes erecta* and *Tagetes patula* appear again. It is remarkable, however, that among the "recovered" *erecta* species there

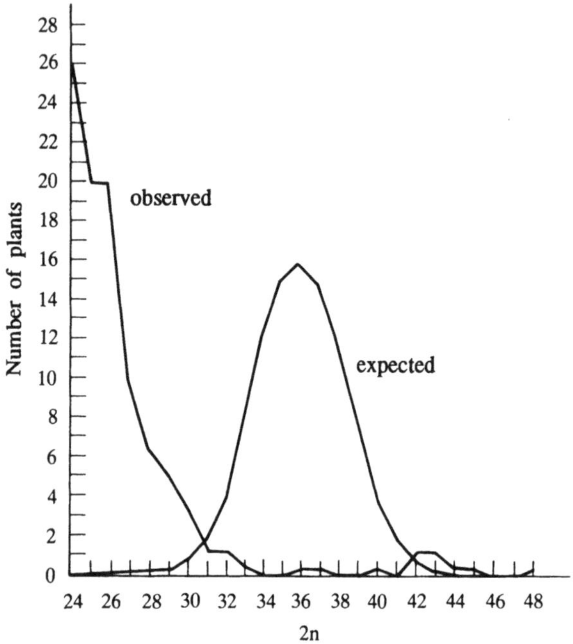

Fig. 47. Expected and observed frequency of chromosome numbers of F_2 plants (Eresen 1966)

are new genotypes, which are no doubt explained by the already mentioned recombination of the chromosomes of the A_p and A_e genome or to a chromatin exchange between the homologous partners of both A genomes, and which may also be of interest for the breeder.

3.6.2 Protoplasts and Somatic Hybridization

In Section 3.4.3, it has already been discussed, for potato, how a genetic problem may be elegantly solved by cell fusion. Since protoplasts can be produced and regenerated into functional plants, the possibility of starting somatic genetics by fusion of these cells without a cell wall has been discussed.

Plant tissue can be digested enzymatically into single cells in a hypertonic (0.6 *M*) mannitol solution with pectinase (ca. 0.1%); by further digestion with cellulase (ca. 1%), the cellulose wall is dissolved within 3–12 h, and protoplasts are obtained (Fig. 48a). Such protoplasts are cultured in hypertonic liquid media until a small callus is formed (Fig. 48c,d). It can then be treated as normal callus and can regenerate functional plants. Such plants are cloned as in rapid propagation systems, but the multiplication rate is much higher; from a given plant in one 10-cm Petri dish, protoplasts can be cultured that will result in 5000 plants. Under the influence of the culture conditions in this system, mutations easily occur

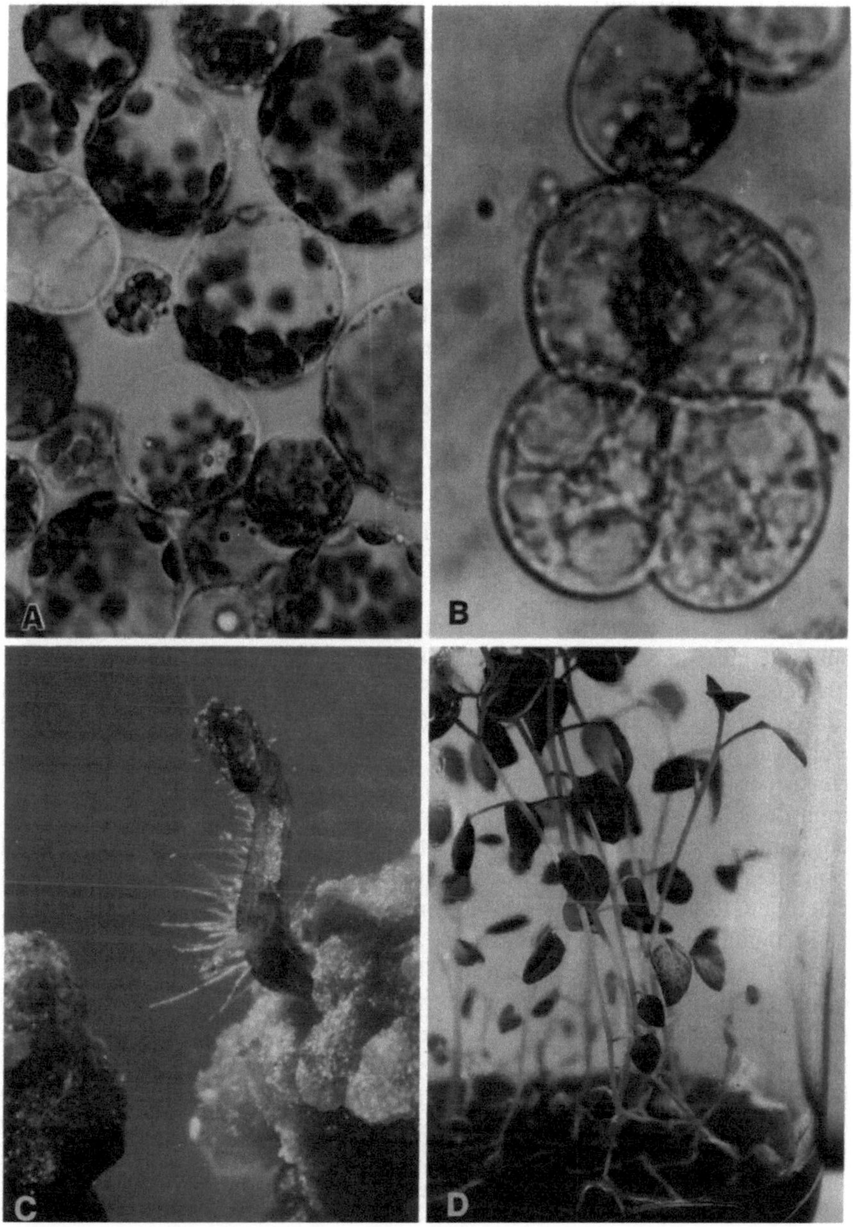

Fig. 48A-D. Plant regeneration from protoplasts. *A* Isolated protoplasts; *B* early division; *C* callus formation and organogenesis; *D* regenerated plantlets (protoclonal variation), resulting in altered protoclones (Sect. 3.1.4). This reduces, under practical aspects, the applicability of such micropropagation strategies.

Today, protoplast regeneration is of practical interest in agriculture for tomato, rapeseed, potato, and alfalfa. In the monocots, a major breakthrough has been achieved with the regeneration of rice plants from protoplasts isolated from cell suspensions in 1985/86, from fertile maize plants and wheat in 1989. The rice plants regenerated from protoplasts have been tested in the field and the regenerants were found to be comparable to the protoplast donor material.

The practical interest in protoplasts as a tool for rapid propagation is rather limited; much more is expected by using them in somatic fusion or in gene transfer experiments. Protoplasts are able to absorb macromolecules like DNA. This is of particular interest in dicots, where vector-mediated DNA transfer is still difficult. Substantial progress in gene transformation programs of cereals demands reproducible protoplast techniques.

Protoplasts not only can be regenerated into functional plants, they can also be fused (Fig. 49b). Such fusions take place spontaneously during protoplast isolation and cell wall degradation, and they can be induced chemically by the polycation of polyethyleneglycol (PEG, 5–25%) in the presence of Ca (50–250 m*M*) at pH 8–10 or via electrofusion, as long as no new cell membrane is formed around the plasma lemma. Under such conditions the protoplasts adhere, and subsequently their membranes, their cytoplasms, and finally the nuclei fuse (for further details, see Gleba and Sytnik 1984). In electrofusion, the protoplasts first form chains in a pulsed electrical field (Fig. 49a). Fusion can be induced by an electrical pulse between 20 and 1000 V, depending on the distance of the electrodes, for 2–100 ms. In potato, the somatic fusion procedure is so reproducible that practical programs have started to use it. There are, however, only limited data available, demonstrating the interaction of genomes and of genomes and plasmons after fusion. Population genetic approaches must clarify whether or not superior clones result from heterokaryons after protoplast fusion.

Somatic fusion allows the asexual addition of two heterozygous genomes without any meiotic recombination. This is very useful when characters are combined that are polygenic and inherited quantitatively. Austin et al. (1985) transferred resistance to potato leaf roll virus (PLRV) and late blight from *Solanum brevidens* into *S. tuberosum* × *S. brevidens* hybrids. Since polyploid hybrids, however, are of no immediate practical interest, it is usually necessary to reduce the ploidy level to the level of the cultivar before protoplast fusion. Furthermore, propagation procedures must be developed that will allow the continuous growth of such a heterozygous genotype without immediate segrega-

Fig. 49A-E. Protoplast fusion. *A* Protoplasts in an electric field; *B* magnification of the protoplast chains formed under influence of the electric pulse; *C* fused protoplasts; regenerated somatic hybrids of potato; *E* tubers from the two dihaploid parental clones (*upper row*), and tetraploid homo- and heterokariotic (*lower row, center*) *hybrids*

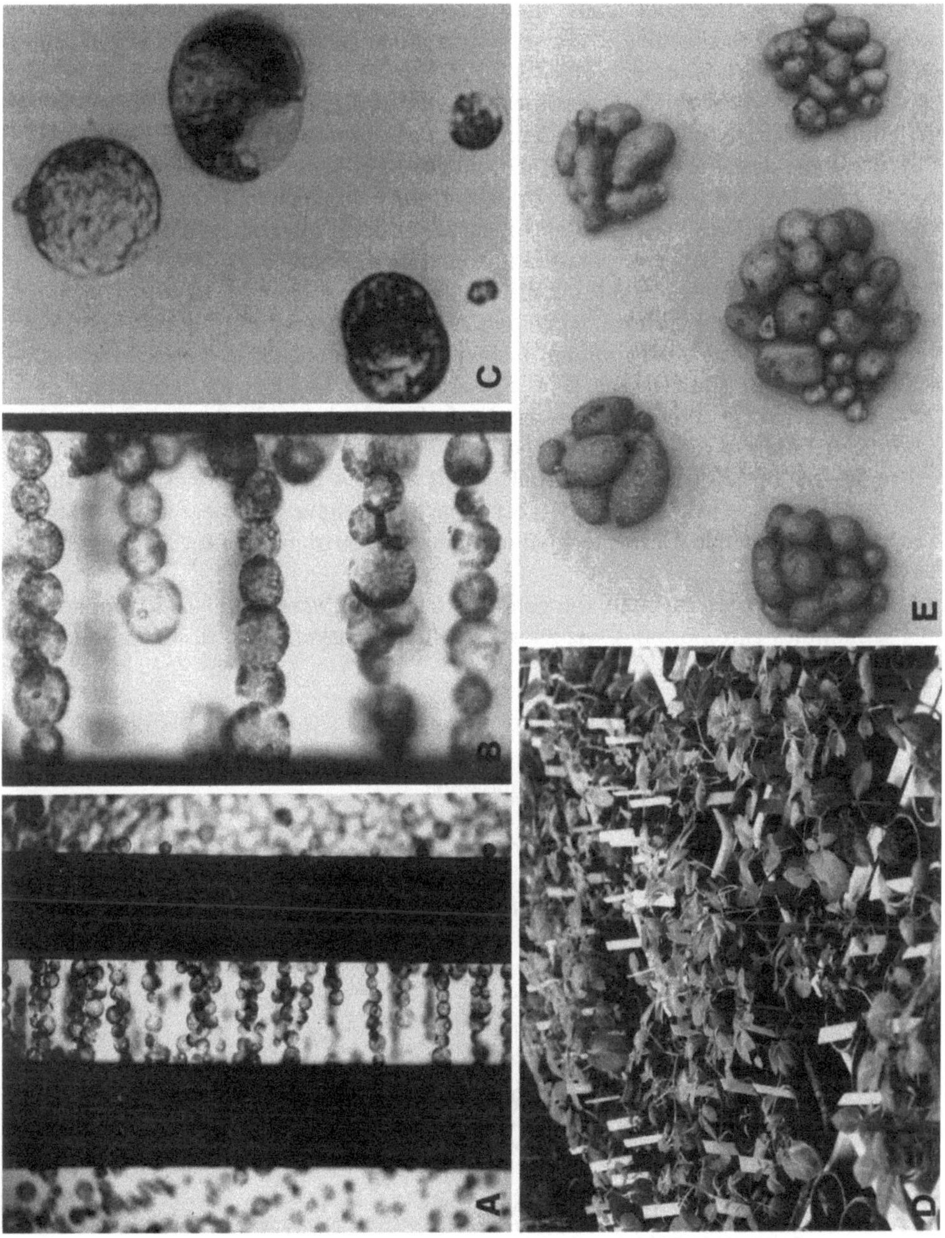

Fig. 49A-E

tion. Among the field crops, these prerequisites are adequately fulfilled in potatoes, since breeding at the dihaploid (2n = 2x = 24) level and vegetative propagation are routine.

It is also desirable to check how far sexual incompatibility, e.g., in combining different species or genera, may be overcome by protoplast fusion. Fusion would allow the direct combination of diploid genomes to result in an amphidiploid hybrid. The classical procedure, by doubling the amphihaploid hybrid to restore fertility, could be omitted. Besides some scientifically interesting hybrids like the fusion product between potato and tomato (Melchers et al. 1978), such amphidiploids are of immediate interest in the genus *Brassica*. In order to investigate the taxonomic limitations regulating compatibility between nuclear and organelle genomes, a program has been started by Glimelius et al. (1990) within the family of Brassicaceae (Fig. 50). Efficient procedures for production of somatic hybrid cells, selection of the fusion products via flow sorting, and regeneration of calli have made possible a rapid screening of different species combinations. Both true and partial hybrids with new unique compositions of the cytoplasm have been produced. Most of the hybrids have some degree of fertility, making it possible for this technique to be used for transfer of foreign genes to our crops.

In this latter example, the genomes that fused were not always complete; sometimes chromosomes were lost, resulting in asymmetric fusion products.

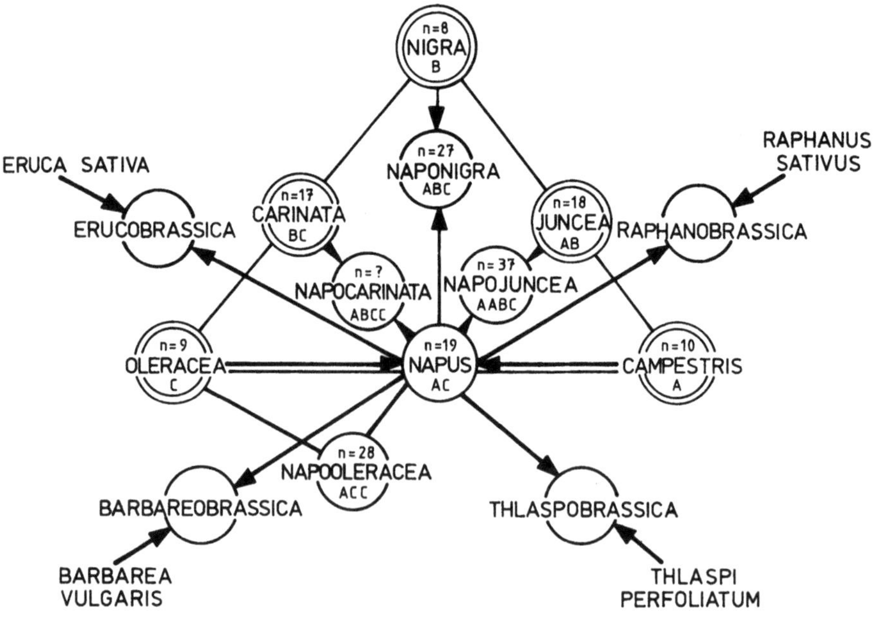

Fig. 50. An overview of different somatic hybridization experiments in the genus *Brassica* and related genera. The *simple circles* indicate newly produced hybrids (Glimelius et al. 1990)

During fusion, only a partial genome is combined with that of the other fusion partner to produce a plant with a few additional characters. The principle idea is that specific parts of a foreign genome are added to a superior genome for its further improvement. The potential use of this procedure was demonstrated, e.g., by transferring the atrazine resistance from a mutant of *Solanum nigrum* to dihaploid clones of *S. tuberosum*. In a similar approach, a chromosome fragment from carrots bearing a gene for resistance to methotrexate could be transferred into tobacco protoplasts. The regenerated plants are similar to tobacco but additionally express the methotrexate resistance. In contrast to sexual recombination, in such asexual procedures the cytoplasms are mixed as well, and allow for recombination of the cytoplasm (cybrids). The asymmetric fusion or cybrid formation may be improved by the donor recipient fusion technique, where one fusion partner is irradiated so as to partially or completely inactivate its nuclear genome.

For applied agriculture, hybrid seed production is of interest. Such programs are started, e.g., in rapeseed and tomato. One problem is the transfer of the cytoplasmic male sterility (cms) allowing for economic production of hybrids. The cms character, missing in *Brassica napus*, was transferred from *Raphanus* by somatic protoplast fusion. Melchers et al. (1990) fused tomato protoplasts treated with 10 m*M* iodoacetamine and *Solanum acaule* protoplasts irradiated with 100 krad X-rays. The regenerated plants had 24 chromosomes and were morphologically identical to the tomato variety but the pollen, although they appeared completely normal, did not germinate. Thus, the tomato became male sterile. For this path towards male sterility the patent is still pending. In asexual fusion, which can be asymmetric or even just form cybrids, the transfer of cytoplasm is achieved; this is normally not transferred by sexual genetics, demonstrating a principle difference between sexual and asexual genetics.

This partial transfer of genetic information builds a bridge to the following section.

3.6.3 Gene Transfer

If one observes the development of systematic plant breeding, it becomes obvious that its efficiency is increasing. In the beginning during selection breeding, already existing genotypes were only chosen for further multiplication; later combination and, increasingly, hybrid breeding offered methods by which traits were brought together in a programmed manner. However, the production of new variability is in classical and also in mutational approaches the result of very accidental recombinations. By using haploids, simpler genomes can be worked with, thus making the combination of quantitative traits more feasible, and in symmetric and asymmetric somatic cell fusion, genomes are added without meiotic segregation, again leading to an increase in predictability of the characteristics of the resulting new generation. In all these techniques, genomes or at least parts of genomes are combined, wherein a desired trait is always accompanied by several undesired characters. Thus, it would be the ultimate aim of breeding for new genotypes to improve a top variety in one single additional character without disturbing the rest of the genome. The new product should be the new variety without any additional breeding steps. Today, we are witnessing the first examples of a successful transfer of specific traits into higher plants by gene transfer. We are, however, still far from being able to construct the super plant by making a new artificial genome. As this area of modern plant breeding is developing very rapidly, here only some principles can be explained; where and to what extent gene transfer will alter the classical and semiclassical breeding techniques will not be evaluated here on the basis of the first spectacular findings.

Techniques that open up to the plant breeder the possibility of transferring in a planned manner characters from one organism to another have been developed in microbial genetics. It should be stressed right at the outset that the expression "gene" has different meanings in agriculture and in molecular biology. Since Mendel's experiments in the late nineteenth century, agriculture has described a gene – the expression was introduced in 1909 by Johannsen – as a segregating character, and has visualized different crossing-over rates as different distances on the Morgan gene map given in Fig. 51 for barley. In plant breeding, a gene is a mathematical expression; nevertheless, linkage groups and crossing-over rates can be symbolized on chromosomes to give a rather concrete picture. In molecular genetics, a gene is known physically as a sequence of nucleotides linked up in the DNA macromolecule (Fig. 8). At the moment, intensive research is being devoted to explaining a gene on the Morgan map as a physical base sequence. The present calculation is that 1 cMorgan roughly equals 10^6 bp. This confirms that we are still far from being able to construct a better plant by use of artificial genes. In order to judge the present knowledge and the future possibilities, some principles of molecular genetics must be explained. This will be done here very briefly, touching on only the principal aspects. For further information, the reader is referred to textbooks on molecular genetics.

We know that a gene or, more precisely, a structural gene, consists of a longer or shorter linear sequence of nucleotides as a part of the huge double-stranded DNA molecule. Together with its promotor – an activating point for an enzyme that transcribes the DNA into RNA, and perhaps some other regulatory mechanisms like DNA-binding proteins (in bacteria we know specific operators) – one or more structural genes form an operon. Such an operon is transcribed into single-stranded messenger RNA (mRNA). At the ribosomes, the mRNA is translated into peptides by linking amino acids, which are delivered from transfer RNAs (tRNA). In all processes, further modifications take place, e.g., processing until the protein(s) form the final enzyme. The enzymes are responsible for the total pathways of the plant. The function and shape of the organism depends on their relative concentrations and activities. With the exploitation of this causality, the connection between genotype and phenotype is successfully established in molecular genetics. This path can be copied to a large extent in vitro in an artificial system.

It is possible to synthesize DNA sequences; it is also possible to transcribe and translate DNA and RNA in vitro and to obtain finally the corresponding

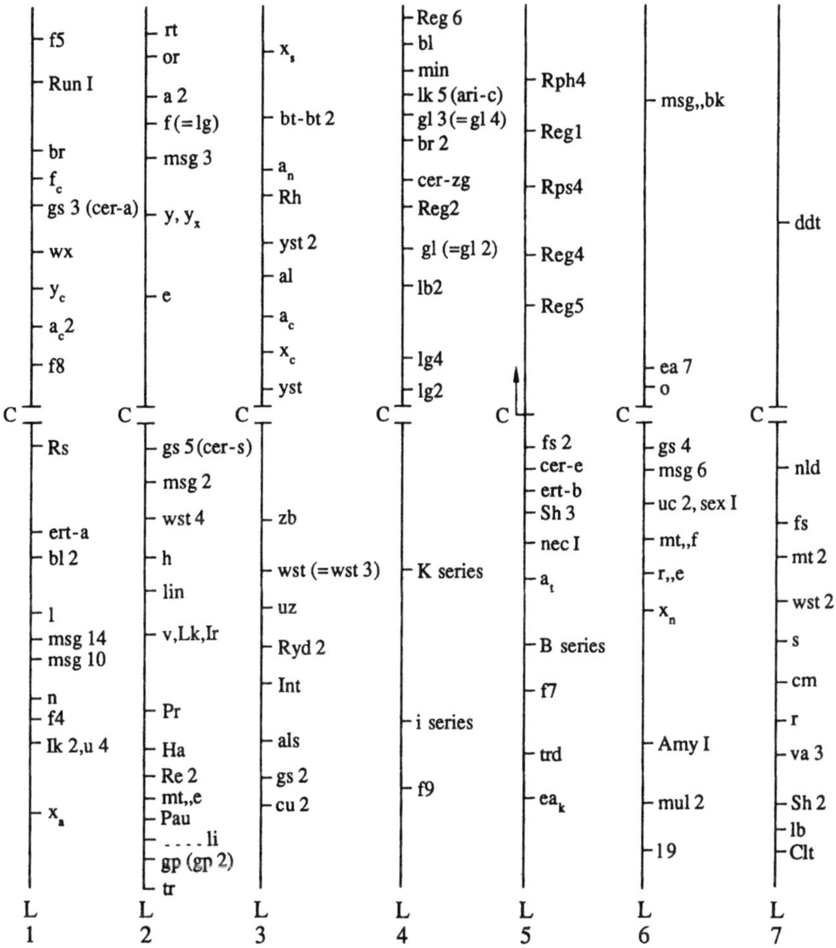

Fig. 51. Morgan gene map of barley (Tsuchiya 1986)

protein. The genetic code (the sequence of nucleotides, three of which code for one amino acid) is unveiled; we do understand the letters of life. But what is even more exciting, this path can increasingly be followed in the reverse direction (Fig. 52). If, for a given phenotype, sufficient quantities of a responsible protein can be isolated, the pathway from the phenotype to the gene can be followed as well; in other words, molecular biologists are starting to construct new genomes.

What is the relevance of this progress for a plant breeder? Can he construct a super plant? Not today! But already now, the following aims are of particular practical interest: production of additional resistances to diseases, environmental

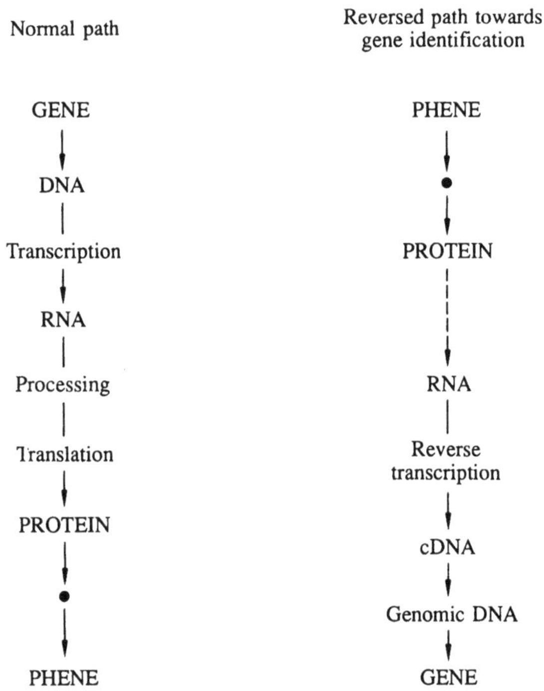

Fig. 52. Path from gene to phene and the reverse as a method of gene technology

stresses, and herbicides; alteration of primary and secondary plant products such as proteins, oils, and carbohydrates; increase of photosynthetic activity; and improvement of symbiosis with soil organisms for a better uptake of nutrients.

To use gene transfer techniques, the following prerequisites must be fulfilled.

1. The gene must be available as a fragment of DNA.
2. This DNA should be clonable.
3. The gene has to be manageable in a transfer system.
4. Integration of the DNA in the cellular genome must be achievable.
5. The transformed cell must be regenerable into a functional plant.
6. The plant has to express the transferred trait.

In principle, all prerequisites are already fulfilled today.

The first structural gene (ribulose bisphosphate carboxylase, rubisco, a part of the fraction-I protein) was isolated in 1977 by Coen et al. Fifty percent of the soluble leaf protein consists of this enzyme, the large subunit protein of which is coded from the DNA of the chloroplasts. Thus, it was an ideal candidate for isolation and identification of its gene. It was followed by the gene identification for storage proteins such as zein, hordein, and phaseolin. In all these cases, for a limited period of time most of the RNA present is responsible for the production of that particular protein. The highest RNA peak should correspond to that protein, thus the RNA was identified. This analytical step from the protein to the RNA is not reversible due to energy reasons and due to its very high complexity; from RNA to DNA, however, the enzyme reverse transcriptase can produce single-stranded copy DNA (cDNA), which can either be enzymatically converted into genomic DNA or, by DNA hybridization techniques, can be probed to cleaved DNA pieces. The cleavage is done again enzymatically by using restriction enzymes, of which about 150 are known today, after the first ones were detected by Arber and Dussiox in 1962. The detection of these enzymes was probably one of the key events of present genetics, which was honored with the Nobel prize in 1978.

Each restriction enzyme cleaves the DNA at different specific points, with or without sticky ends. The small pieces can be analyzed for their nucleotides and, like a puzzle, several pieces can be linked to end up in a nucleotide sequence of a part of a gene or of a complete gene.

Having identified the corresponding DNA of a protein, this DNA has to be propagated, it has to be identically cloned to result in enough DNA for transfer experiments. For cloning, bacteria are used. First, the DNA is linked to a plasmid, cosmid, or similar vectors that are able to replicate. Such circular DNA is quite common and is known by abbreviations (often the initials of the responsible scientist or laboratory) such as pBR 322, one plasmid that was used in a number of fundamental transformation experiments. Once the plasmid with its additional information is in a bacterium, it will be propagated; normally, some additional sequences, such as an antibiotic resistance, are added that allow the easy detection of bacteria which propagate the engineered plasmid with its foreign piece of DNA. Today, cloning of a gene is no longer a problem.

For the subsequent gene transfer, three methods are available:

1. microinjection,
2. insertion via microprojectiles,
3. direct transfer,
4. vector-mediated gene transfer.

Based on the positive results of microinjection in animal systems, this technique was tried in plants. Although under applied aspects such a laborious method will only be of very limited importance, one procedure will be discussed here. Neuhaus et al. (1987) microinjected gene constructs into multicellular structures of *Brassica napus*, which have a high competence for plant regeneration through embryogenesis. Microspore-derived embryoids of *B. napus* were individually microinjected with the neomycin phosphotransferase II (NPT II) gene, under control of the cauliflower mosaic virus promotor 35S and the transcription terminator of a modified pDH 51. Transformation efficiencies between 27 and 51% were determined in primary regenerants. Stable integration of full-length, microinjected genes into high-molecular-weight DNA was proven by Southern blot analysis of genomic DNA isolated from regenerated plants. As the starting material was multicellular, chimeras were produced. The chimeric nature of such primary regenerants was overcome by in vitro segregation through secondary embryogenesis, resulting in wild-type and pure transformants. Today, experiments are in progress to transfer this system to embryogenic, microspore-derived cultures of barley and wheat.

As an extension of the microinjection with syringes, high voltage microprojectiles or air pressure guns with microprojectiles may be used. Laser beams are also in development.

The basic principle of the direct gene transfer method has been summarized by Potrykus et al. (1987). In this procedure, foreign DNA is linked to a strong promotor and to marker sequences, and is incubated together with protoplasts. Under the influence of 8% PEG, heat shock (5 min at 45 °C) and/or electric shock (electroporation, 1400 V, 10 µs), the DNA is taken up and may be integrated into the nuclear DNA. It could be demonstrated that using direct gene transfer, a number of monocots could be transformed (Table 20).

Vectors are most commonly used for gene transfer, particularly the Ti-plasmid of *Agrobacterium tumefaciens*. For technical details, the reader is referred to the review of Horsch et al. (1987). For *A. tumefaciens*-based gene transfer, the infection of a plant cell by the bacterium is a prerequisite. This occurs easily in nature only with dicots, while monocots are no host. Thus, this transfer technique is hampered by host-range limitations, which are especially disturbing when aiming at gene transfer into cereals. The host range poses an even more severe problem in the other biological vector system, plant DNA viruses. Despite its restriction in host range, *Agrobacterium*-mediated transfer has gained in practical importance. In the following section, examples will be described where application in agriculture is expected soon. The basic principal is illustrated in Fig. 53.

Most progress has been obtained in incorporating specific tolerance to herbicides, and plants with tolerance to nonselective herbicides will be the first practical results of plant gene technology. One reason for the rapid advances in this field is the fact that the biochemical pathways of modern herbicides are known; they are quite often based on just one or a few enzymes, allowing for a straightforward strategy.

Thus, a specific tolerance to one of the most commonly used herbicides, glyphosate, has been transferred first to petunia and subsequently to tobacco and tomato. The strategy used was an overexpression of 5-enolpyruvylshikimate-e-phosphate synthase (EPSPS), which is blocked at normal concentrations by the herbicide. Overproducing mutants were first screened for in-cell suspension

Table 20. First transformation experiments in monocots

Species	Transformed gene construct	Reference
Lolium multiflorum	NPT II	Potrykus et al. (1985)
Triticum monococcum	NPT II	Lörz et al. (1985)
Zea mays	CAT	Fromm et al. (1986)
Oryza sativa	CAT	Uchimiya et al. (1986)
Sorghum	CAT	Ou-Lee et al. (1986)
Panicum maximum	CAT	Hauptman et al. (1988)

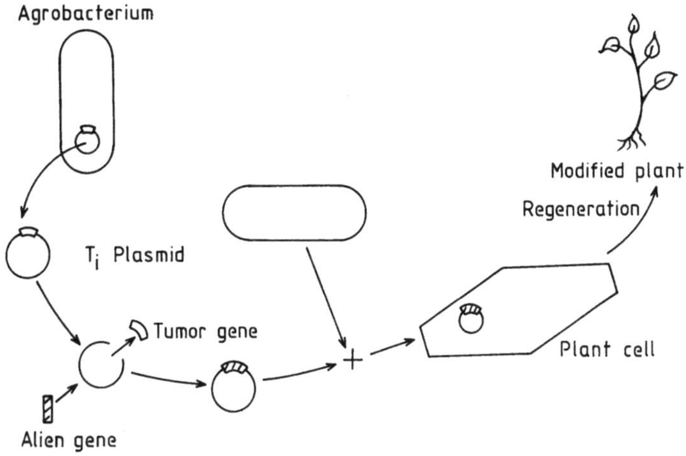

Fig. 53. Transfer of alien genetic material into plants using *Agrobacterium* as vector

cultures of petunia by sequential step-up from 0.1 m*M* glyphosate to 20 m*M* glyphosate. In the tolerant cultures, about 20 times more EPSPS was produced. The pure protein was then partially sequenced. From there, the mRNA could be used to isolate a cDNA, which was finally transferred via *Agrobacterium* to petunia cells. Transformed callus proved to be tolerant to high glyphosate concentrations.

The objective of genetically engineering plants with herbicide tolerance is philosophically different from that of disease resistance: one is intended to permit the continued use of pesticides and the other to use less pesticides. Unfortunately, our knowledge of biochemical mechanisms of disease resistance is very limited, and a straightforward strategy comparable to the approach in herbicide resistance is not yet available. Nevertheless, substantial progress has been made in inducing virus resistance via gene technology. This strategy, however, is quite different from the previous one. It is known that a virus causes cross-protection or praemunity, which means that a related virus cannot infect the same plant. The coat protein of the virus is responsible for this protection (Fig. 54). The gene responsible for the coat protein production of TMV via *Agrobacterium* has been transferred to tobacco. It was expressed and prevented symptom formation in tobacco plants after TMV inoculation. The same strategy was successfully used to protect potatoes against viruses.

Another promising approach is the incorporation of the *Bacillus thuringiensis* toxin gene into host plants to ward against insects. The toxin, a protein, is produced very specifically in *Bacillus thuringiensis*, a natural antagonist to caterpillars, diptera, and beetles. The gene for the toxic protein has been isolated, cloned, and transferred to tobacco and some crop plants. In tobacco, it has been demonstrated that the gene is expressed and the correct gene product is formed. After some amplifications, the concentration of the toxin is high enough to protect the plant against attack by caterpillars.

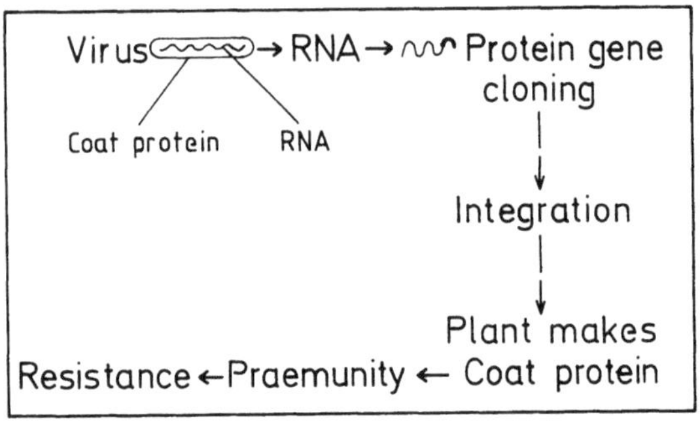

Fig. 54. Diagram of the induction of virus resistance via gene technology

The situation in the area of environmental stress resistance will also briefly be mentioned here. The strategy to influence physiological parameters or secondary products is straightforward. Seeds are quite often an important nutrient because of specific amino acid compositions. Experiments to influence the protein or starch content of the potato tuber are under investigation (Table 21). Strong interest exists in clarifying the pathways of fatty acids to influence the oil and lipid composition of *Brassica*. These approaches are of particular interest, as nature deposits in the oleosomes of seeds a wide range of very different C18 triglycerides, thus enabling analysis of specific gene functions and transfer of such characters (detected, e.g., in wild species) to crops. In cases where just a few enzymes are involved in specific product formation, gene technology has great potential. It should soon be possible to identify and transfer genetic information for one enzyme, while it will probably take quite a while before complex characters like most disease resistances can be handled.

Some strategies also demand the blocking of active genes. Here, the approach using antisense DNA is promising (Fig. 55). Although the action is not yet completely understood, the principle function of this method has been shown in the starch pathway of potato.

Table 21. Change of the protein content of the potato tuber by transformation with antisense DNA (Sanchez-Serrano et al. 1990)

Tuber protein of potato		2%
of which patatine	ca. 40%	
reduced via antisense RNA to	4%	
Total protein content after transformation		1%

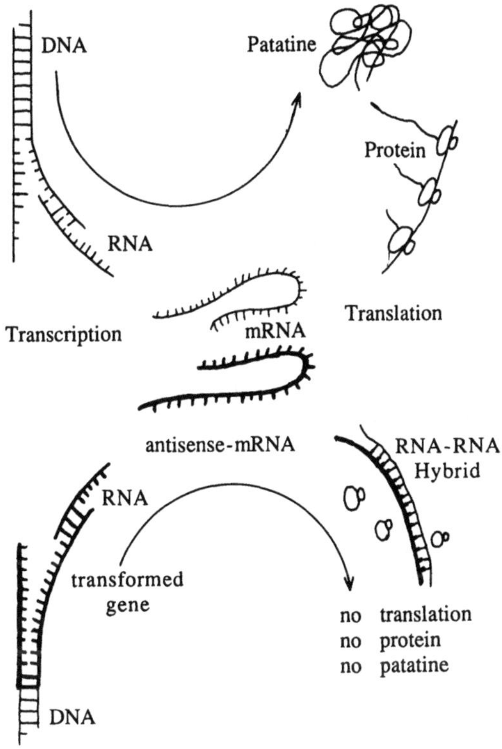

Fig. 55 Hypothesis for the function of the system of stopping gene functions via the transfer of antisense DNA

3.7 Special Selection Procedures

It has already been emphasized in Section 2 that not only is selection carried out in selection breeding, but also in combination and hybrid breeding. The smaller the possibility of the occurrence of desired genotypes in a natural population in the generation following a cross or after mutagenic treatment, the greater the number of individuals or progeny that must be examined. The breeder is therefore interested in the development of methods that allow the desired individuals to be determined quickly and easily. Such *special mass selection procedures* should, as far as possible, be applied to a plant in an early phase of its development in order to reduce breeding costs; they therefore presuppose an early diagnosis. In general, these would be carried out on field crops, whole plants, or parts of plants (in vivo). Recently, it was possible to develop methods which allowed the recognition of species traits on plants, tissues, or organs cultivated on synthetic media in vitro.

3.7.1 Early Selection in Vivo

Mass selection and early diagnosis are occasionally simultaneously effective in a method, as can be seen in the classic example of the breeding of alkaloid-free (sweet) lupins by von Sengbusch (1942). The basis for this breeding is formed by the reaction of iodine-potassium iodide: in the presence of alkaloids a change in color can easily be recognized in seeds and leaves; in the absence of alkaloids such a color change does not take place. For example, if one removes leaves from a young plant in such a way that one also removes a piece of the epidermis from the stalk, these will change in color when a plant containing bitter constituents is dipped into an iodine-potassium iodide solution. The color change is absent in alkaloid-free leaves. The next step in the breeding procedure consists of a rational organization of all the steps that allow the testing of lupin field crops in a short time in the early stages of development. The fact that only one out of ca 36 000 examined plants in a cross-section of a utilized population was found to be free of bitter constituents shows the need for the development of rapid methods as a prerequisite for breeding success.

An early selection with the help of chemical reactions led to the discovery of forms low in cumarin in the case of sweet clover, and to bitter-constituent-free forms in the case of cucumbers. Due to the development of improved chemico-physical laboratory procedures for determining fatty acids, which depend upon chromotographic separation processes, it was possible to make great progress in the quality breeding of rapeseed.

Duè to its content of eruca acid, rapeseed oil is not equivalent to soy, sunflower, or peanut oils and therefore can only be used in a very limited amount in the production of margarine. The discovery of a mutant free of eruca acid in "Liho" provided the impetus to free winter rape cultivated in Germany of eruca acid by means of combination breeding. The selection can already take place on the seeds developed from F_1 plants since with the improved method the acid pattern can already be determined in the embryos of these F_2 seeds. After a 24 h soaking, the embryo is freed from the episperm and the outer cotyledon is prepared for the determination of fatty acid. The rest of the seed along with cotyledon rootlet, and shoot cone develops on damp blotting paper into a plant that is either kept or rejected, depending on the results of the investigation.

Mass selection procedures combined with early diagnosis have obtained a special importance in the field of resistance breeding, where already young plants can be tested for their reaction to plant and animal parasites as well as to viruses. These procedures are dealt with in Section 4.3.

Early diagnosis can depend upon the effectiveness of genes that already manifest themselves in the early phases of a plant's development. For example, it is possible to determine whether a snapdragon will form red or white blossoms from the presence or absence of color on the underside of the first leaves of a seedling. The type of red coloring, eosin or fuchsin, can also be determined in this early phase. Genes with pleiotropic effects can also be successfully used in early diagnosis. The mutant "Weiko" (white grained) appearing in the yellow lupin is not only distinguished by a white episperm, but also by a light-yellow blossom, higher protein content, and low sensitivity to alkaline soil. The use of pleiotropy in the selection rests on the assumption that a complex of characters is controlled by one or several closely linked genes. By selecting an easily recognizable partial characteristic of a complex, e.g., the white grain color of the "Weiko" mutant, a simultaneous selection follows based on high protein content, a character that could only be directly determined by chemical methods. Complex mutants, which exhibit a monohybrid segregation after crossing with the original form, often exhibit an independent variation of the partial characteristics after they are imbedded in a different genetic background.

Furthermore, it has been shown in the quality breeding of corn that the usefulness of pleiotropy for special selection procedures, including fast methods and early diagnosis, is limited. In the case of corn, the breeding of strains with biologically high-value protein depends upon the determination of two gene mutations for high lysine content: opaque-2 (o-2) and high tryptophan content; floury-2 (fl-2). The selection of lysine-rich genotypes from crosses with flint and dent forms is carried out relatively easily due to the pleiotropic effect of the opaque gene, since the lysine-rich genotypes are recognizable on the endosperm. After several backcrosses, one also has lysine-rich grains with transparent and translucent endosperm. The pleiotropic effect of the opaque gene is lost and therewith the possibility of indirect visual selection for high lysine content, so that for this selection a special chemical method had to be developed.

Linkage of an indicator gene with a doubling gene is also used for early diagnosis in the breeding of gillyflower. Double-flowering gillyflowers are completely sterile in both sexes. Their propagation results from single-flowering plants of certain tribes, which are characterized by progeny that segregate 50% into single-flowering and 50% into double-flowering plants. The permanent heterozygosity and the deviant segregation 1 single flowering: 1 double flowering as opposed to 3 single flowering: 1 double flowering are due to the special constitution of the single-flowering plants of this type. They are heterozygous for the gene S-s, which is linked to the lethal gene Ll in the following way: $\frac{Sl}{sL}$. The gene l has a lethal effect only on male gametophytes, so that Sl pollen cannot be produced. The 1:1 segregation of the single-flowering plants $\frac{Sl}{sL}$ is evident from the following schema:

♀ ♂	Sl lethal	sL
Sl	–	$\frac{Sl}{sL}$ = single flowering
sL	–	$\frac{sL}{sL}$ = double flowering

One sees further from this segregation that never more than 50% of the double-flowering types can be anticipated. This was avoided when a monogenically segregating mutation Cc for normal and light-green leaf color was found, which is linked to the doubling gene Ss, and was built into the permanently segregating heterozygotes. Fifty percent of the heterozygotes SlC/sLc segregate into 50% single-flowering plants with dark-green leaves and 50% into double-flowering plants with light-green leaves. If only light-green plants are planted, the result is a unified crop of exclusively double-flowering plants, as long as the exchange occurring often between the genes S and C has not led to the combinations SLC.

To speed up the breeding process, especially for types with generation spans of several years, e.g., in the case of pome fruit and stone fruit and a few ornamental plants such as tulips, hyacinths, etc., early diagnosis functions as a useful tool in the selection of segregating generations. The prerequisite here is that characters in the early phase of development must be correlated with those in the generative phase. Early rejection of useless genotypes from the population allows a great reduction in breeding costs and intensifies the selection. For example, in the seedling progeny of tulip crosses the mean bulb weights between different ages are so correlated that already in the 2nd year of growth a preliminary selection with respect to this important characteristic for every tulip species, a rational multiplication can be made, while the selection of blossom characteristics, etc. is not possible before the 6th year of seedlings.

Although selection in the first generation after the cross is also possible in species with generation spans of several years as a result of heterozygous parents (see Sect. 2.1.4), nonetheless it cannot be determined whether or not the desired variation has been achieved for several years. The breeding of the selection generation is associated with very high costs. One can greatly reduce the risk of cultivating excess useless material by selecting parents with an appropriate hereditary constitution. Therefore, especially in the case of breeding for quantitatively inherited characteristics, only such genotypes should be used as parents which contain as many heritable components of the characters as possible. Since the phenotypic variance of characters in a segregating generation is composed of the sum of genotypic and environmentally conditioned variance ($V_P = V_G + V_U$), the proportion of genotypic variance can be determined by the quotient $V_G/V_P = h^2$ (also H). In a broad sense, this serves as a measure for the heritability of a quantitative character (h^2 = heritability). The estimation of genotypic variance is generally determined by parent-progeny testing, i.e., procedures that are time-consuming and costly. If clone types, provided as cross parents, are tested in various areas over a period of several years, then one can estimate the environmental variance of the quantitative characters of interest. If this environmental variance is subtracted from the total phenotypic variance, one has an estimated value for the genotypic variance of $V_G = V_P - V_U$ and therefore also for h^2.

A further valuable tool for the selection of suitable cross parents is the determination of *ecovalence*. This parameter, is a quantitative measure for the evaluation of ecological adaptability. It derives from an expanded variance analysis of scattered testing, in that the totality of all interaction components between genotypes and environmental factors are divided into each individual trial member. The smaller its proportion on a genotype, the greater is its ecovalence.

The use of biometric parameters speeds up and intensifies the breeding process not only by the selection of suitable cross parents, but also by more effective selection in the segregating generations. Before starting the selection from diverse populations, it should be known whether the characters to be selected in these populations indicate a genetic variance and, if so, to what extent. The selection advantage (R) to be anticipated from a selection can be estimated by the formula $R = i \cdot \sigma_p \cdot h^2$. In this case i is a parameter of the intensity of selection (it can be determined from pertinent tables) and σp is the phenotypic standard deviation of the original population. Since in the case of the dominant effect of the allele in the progeny of selected heterozygous genotypes, recessive homozygotes are again segregated out and thus reduce the selection success, the parameter h^2 in the above-mentioned formula (h^2 defined as heritability in a broader sense) cannot be used without qualification. Since heterozygotes cannot be maintained in self-fertilizing species, and in cross-fertilizers only to a limited extent, the dominance effects in such structures generally cannot be used. Occasionally, the so-called additive portion of the genetic variance (V_A) is completely usable. There are several available procedures which allow the extent of the additive variance components to be determined as well as the degree of the nonadditive portions. From V_A one is able to determine via V_A/V_P heritability in the strict sense, which henceforth is to be included in the formula for selection advantage see Section 2.1.1.2.

In the relationships between two characters or between different developmental phases of a character included in the breeding experiments, it is a case of *phenotypic* correlations. These contain a genotypic and an environmentally dependent component. We must take into consideration linkages of genes and pleiotropy as genetic causes for correlation. Also, in the case of closely linked genes one must reckon with crossing-over, which leads to a recombination of genes and thus also leads to a changed production of characteristics. Moreover, the pleiotropic effect of a gene can be dissipated through storage in a different genetic background (see above). The modifying changes in quantitative characters through environmental factors will occur particularly in the case of characters with low grades of heritability. A correlation determined in a particular population therefore is not universally valid and cannot be applied in the selection of other populations and under other environmental conditions, so that a preceding critical examination of the genetic conditions of the population in question as well as the environmental factors is essential. In addition, in almost every population "correlation breakers" are found, and in lines or families selected from a population many changed correlation coefficients can be determined.

Correlations are not necessarily always genetically determined; they can also depend on the interaction of a physiological process, so that several characters are functionally dependent upon one another. Tetraploidally, induced cell enlargement and the simultaneous changes in the size ratio between nucleus and cell volume has far-reaching consequences for the development of the plant in terms of morphology and physiology. For example, the lower osmotic values in the enlarged cells of rape cause a reduced winter hardiness; disturbances in meiosis caused by production of multivalents lead to fertility problems and thus to reduced seed yield, etc. For example, if one is able to find a microcellular mutant in a tetraploid population, one would also, in all probability, determine changes in other characters and characteristics that are physiologically dependent upon cell size and thus correlated with it. These physiological correlations, which were labeled by Barthelmess (1964) as genetic correlations in a broad sense, are not (or at least much less easily) changeable and are therefore better suited for the early diagnosis or selection of secondary characters.

In any case, correlations should only be used as tools in selection after very careful testing, especially when causal links between the characters are not recognizable in a biological sense. If the breeder relies upon a correlation that does *not* actually exist in his material, there can be no breeding success. The error is greater still when the breeder does not make use of an *actually* existing correlation but uses conventional methods, which take longer, but are with certainty effective.

As already mentioned, the phenotypic correlation coefficient, due to its environmental dependence, is not a good index for estimating the success of an early selection. However, for this, the genetic correlation coefficient is of a suitable size that can be derived from a covariance analysis. If one labels the character upon which selection is performed as x and the dependent character y, then the selection advantage to be anticipated from such an indirect selection $CR = i \cdot \sigma_{py} \cdot h_x \cdot h_y \cdot r_A$, whereby

σ_{py} = phenotypic standard variation of the character y;
r_A = genetic correlation of the characters x and y;
h_x and h_y = in each case, the root of the heritability values (in the strict sense) of the characters x and y.

The indirect selection is, however, only more successful than direct selection if $h_y \cdot r_a > h_x$ (Franke and Fuchs 1980). Prerequisite for the application of this parameter is that the characters exhibit a normal distribution, that the pairs of characters scatter in a straight line, and that the variance in all areas of the straight line be approximately equal.

From the wealth of possible applications of biometric methods in plant breeding, only a few examples of its importance have been illustrated here, without touching on the development and practical applications of these parameters. In conclusion, the most important areas are summarized, in which the interpretation and use of estimated values of heritability and genetic variance might give the breeder an answer to the following questions (Dudlay and Moll 1969):

1. In existing populations, synthetic species, and cross-fertilized species is such a large genetic variation available that improvements of important characters can be achieved by selection?
2. To what extent must the breeding material be checked in terms of years, locations, and repetitions in order to recognize superior populations or genotypes?
3. Which genetic population has as a basis for improved breeding material the best chances for success?

4. Which breeding procedures lead most quickly and effectively to the goal?
5. What type of strain is worth pursuing under the given conditions: a hybrid species, a synthetic species, a pure line, or a species to be reproduced by panmixis?

The biometric procedures in plant breeding rely above all on the insights and methods of population genetics and quantitative genetics. The interested reader is referred to the literature cited.

3.7.2 Selection in Vitro

Conventional programs are already using leaf pieces for testing defense mechanisms against fungi (Fig. 56). Because large numbers of individuals can be screened in a small space, cell culture allows the use of microbial procedures for selection. The first experiments were performed with callus cultures; suspensions and single-cell cultures were used later for more precise experiments.

With regard to the screening of calli for resistance, Morel (1946) realized, for the system *Vitis/Plasmopara*, the possible advantage of growing pathogens in vitro together with their hosts. However, the further development of dual culture techniques has not been very rapid. A dual culture of either haploid cell suspension or stem-embryogenic cultures of rapeseed was used, together with *Plasmodiophora*, as a general screening system. Rape cell suspensions and *Phoma* spores (10^4–10^6) were grown together in liquid media and then transferred as a mixture to agar plates. However, there was no relationship between the reaction of a plant to pathogens and tissue cultures, and the reaction of the same genotype as a plant or vice versa, that is, the rapeseed calli "resistant" in vitro did not give rise to resistant plants. A similar conclusion was drawn by Vardi et al. (1986), who found that culture filtrates of *Phytophthora citrophthora* cannot be used in citrus breeding as a selective tool in vitro.

In contrast to these negative reports, Helgeson et al. (1972) proved that tobacco calli resistant to *Phytophthora* could be obtained from resistant plants, and that susceptible plants yielded only susceptible calli; the disease expression in vitro appeared to be closely related to expression at the whole plant level. Buiatti et al. (1985) found a correlation between in vivo resistance and in vitro response of *Dianthus*, not only to *Fusarium*, but also to elicitors of the fungus. A number of experiments were done with crude culture filtrates of the pathogens, e.g., in potato with culture filtrates of *Phytophthora*, and Sacristan (1985) reported on screening in the *Brassica/Phoma* system. Not all regenerants from selected rapeseed cultures showed increased resistance, but the proportion of resistant or tolerant plants was up to six times higher among regenerants from selected cultures than among plants from controls. The first three progeny generations of rapeseed were tested for their level of resistance; unexpectedly, segregation was found in the progenies of plants classified as resistant, although in all cases the proportion of resistant plants was higher for progenies obtained from resistant material, indicating genetically defined resistance. It became

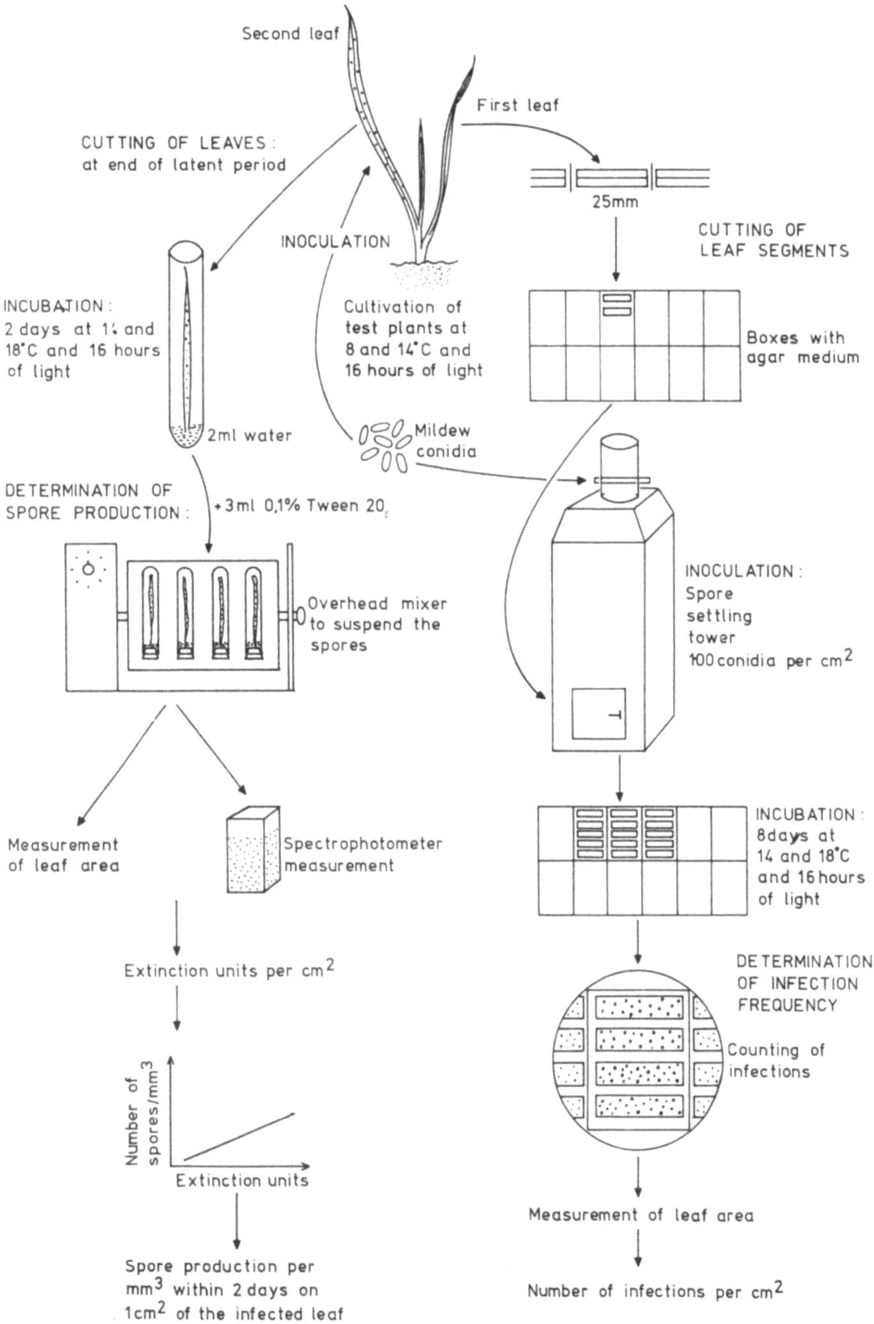

Fig. 56. The test system for the quantitative assessment of mildew resistance using the two components spore production and number of infections (Lind 1983)

evident, however, in several experiments – with and without in vitro selection – that variation in characters, besides the one aimed for, occurred in vitro, mostly in an undesired direction.

The first experiment where a controlled selection for resistance to a pathogen was made with the responsible toxin alone was reported for maize callus. The reversion through somaclone formation from susceptibility to resistance to *Helminthosporium maydis* was, however, associated with genetic changes in the cytoplasm leading to reversion from male sterility to male fertility. Thus, this result was of no practical importance.

As already stated in Section 3.1.4, work with single cells would increase the chance of detecting differences normally covered by cross-feeding. Because the reproducibility in potato protoplast regeneration has been increased, experiments could be started using the crude exotoxin to screen potato protoplast suspensions for *Fusarium coeruleum* and *F. sulphureum*. The relative concentration of toxins was directly correlated with the inhibition of the respiration activity. Due to the rapid passage through the callus level, after regeneration all clones looked rather phenotypically uniform and were fertile. Retesting of leaves or secondary callus revealed a significantly increased stability against the *Fusarium* toxins. Three years of field tests with these potato clones expressed visible differences between the selected resistant plants and the unselected control one; the differences could not be made significant, however, due to very strong environmental influence.

Besides disease resistance, somaclones have been used for detecting specific herbicide tolerances. Here a rather high success rate has been obtained.

The contrasting results, particularly in the experiment aimed at disease resistance, demonstrate that in vitro screening cannot become a general procedure for a huge number of characters. To what extent correlations or causal dependence exists between an in vitro reaction and the phenotype of the whole plant must be checked first. The relationship between reactions of calli with reactions of the same genotypes as complete plants can be very close when the toxic compound responsible for the virulence is used to detect sensitive individuals at an early developmental stage. In all such in vitro selection systems, it is quite common to observe habituation. This means that the resistance or tolerance is present only as long as the selective agent is present. The desired new trait is not fixed in the genome or plastome but only epigenetically expressed. Before a stable new clone can be considered, further generations must be checked.

Plants are not only used to feed man and animal but are also used for the production of nonfood compounds. Of particular importance are primary products of the biochemical pathways: oil, fat, protein, and starch but also fibers and wood. While chances of increasing the production of such compounds via in vitro selection are today rather limited, it is expected that in vitro selection might contribute to the area of secondary plant product formation. Plant cell cultures might have economic potential, particularly for pharmaceutical products where a chemical synthesis is still not possible. An in vitro production coupled to in vitro selection for high producers would have the following advantages.

1. As pharmaceutical plants are grown predominantly in tropical or subtropical countries or must even be collected as wild herbs, industry would become insulated from problems of importing that may arise after poor harvests or other unforeseen problems.
2. The products can be produced in unlimited quantities or their amount can be easily adapted to the specific demands.
3. Secondary compounds produced in vitro under completely controlled conditions would guarantee a high uniformity.

Secondary plant products are formed and stored in special organs of the plant. The question as to whether such a production can be applied to dedifferentiated cell cultures, e.g., in suspension cultures, can be answered principally by "yes". However, there are still more negative than positive examples of the reproducible formation of secondary products in vitro. Normally during culture, special producer cells are selected that are no longer capable of regenerating. This is not an obstacle, however, as for the most part, no regeneration of functional plants is desired from such cells.

The tropical plant *Morinda citrifolia* produces anthrachinones in its roots. It was the first successful example for the production of this compound in cell suspension cultures (Zenk et al. 1977). This example, however, had no practical importance. One of the most prominent applications of this technique is the production of shikonin in *Lithospermum* cell cultures. The complete plant grows very slowly, thus increasing the price for this compound, which is used in pharmaceutical and cosmetic products, as well as in the textile industry. Cell-culture-derived shikonin became well known when lipsticks containing shikonin as dye appeared on the market in Japan as "Biosticks."

Aside from this curiosity, *Digitalis lanata* is mentioned as a second example. From this species two main alkaloids are isolated in very similar amounts: digoxin and digitoxin. Both are used as a heart medicine, but there is much stronger need for digoxin. It became possible to modify the complicated molecule digitoxin into digoxin by the use of *Digitalis* cell suspension cultures, to scale up this process to large volumes and to make it economical (Alfermann et al. 1985). Industry, however, is still reluctant to change the present production process and to adopt this new procedure. The situation is similar in cultures of the medicinal plant *Catharanthus roseus*, where the in vitro steps were successful but an economic realization has not yet taken place.

Because plant cell cultures tend to clump, scaling up from small 1–5 l vessels to large fermenter volumes is always very difficult. Furthermore, cell cultures grow rather slowly compared to bacterial cultures, and the media are rather expensive. Thus, there is a trend today towards isolating the enzymes responsible for the specific compound, in order to find the coding gene sequences and transfering the cloned gene into bacteria. This would enable industry to use well-established bacterial culture procedures. Similarly, it is expected that via gene technology (Sect. 3.6.3) the amount and quality of primary nonfood products can be increased.

3.7.3 Use of Molecular Markers

Modern molecular genetics not only offers techniques for gene transfer (see Sect. 3.6.3), it is also a very powerful tool in selection. Accurate diagnosis of pathogens must precede each plant treatment and each attempt to breed for agronomic traits. Morphological and biological tests (phenotypic level), chromosome identification (chromosome level), and immunological procedures (gene product level) are of utmost importance in this respect. Molecular understanding has opened up the possibilities of monoclonal antibodies. Still, none of these techniques, although already on the protein and responsible enzyme level (gene product level), deal with the causal agent, the DNA, or differences in the DNA (DNA level). Direct DNA diagnosis allows the immediate detection of alterations without uncovering the effects of an alteration.

Thus, we have today three systems, besides cytological chromosome banding techniques, to develop molecular markers providing additional information on the phenotypic appearance of a plant: (1) isozyme patterns; (2) RFLP maps; and (3) gene probes.

The electrophoretic procedures for isozyme diagnoses are already routine in variety assessment. For potato, an extensive catalogue exists giving electropherograms for a wide number of cultivars (Stegemann and Schnick 1985), and also for primitive cultivars and wild species. Furthermore, specific enzymes or enzyme patterns could be used as markers for resistances, such as the endopeptidases for resistance to *Pseudocercosporella* or esterase for resistance to barley yellow mosaic virus.

Isozyme analysis is completed by maps obtained after cutting the DNA with restriction enzymes and hybridizing to DNA probes. The maps of hybridizing DNA fragments appearing after migration in an electrical field and expressing the restriction fragment length polymorphism (RFLP) represent an important tool in describing the variability in crop plants (Beckman and Soller 1986). This is due to their large diffusion in breeding and genetic resources, which permits saturation of the genetic maps for a given species. These are free of secondary effects on economic characters, effects which, due to their pleiotropic action, are frequently associated with the segregation of alleles affecting morphophysiological traits (Table 22). Figure 57 gives a theoretical example of how, by the use of RFLP markers, a desired character can be detected.

Finding RFLPs is dependent on the number of probes available. The number of probes is steadily increasing, and for maize already some RFLP kits containing a total of ca. 1000 probes are commercially used. There is no immediate need to always find the genome part directly responsible for a character expressed at the phenotypic level. It would be sufficient for applied breeding strategies to detect characteristic RFLPs correlated to phenotypes. Correlations are of the utmost importance, particularly from the standpoint that the agriculturally important characters are not coded just by one gene but are of polygenic origin. In tomato, introgressed chromosomal segments can increase soluble solids in the fruit. Two of the introgressed segments have been identified with RFLP markers and localized to the end of chromosome 10 (Tanksley and Hewitt 1988). It has thus

Table 22. Advantages of the use of RFLPs for genome characterization

1. Present everywhere
2. Mendelian inheritance
3. Codominant expression
4. No pleiotropic effects
5. Independent of the environment
6. Present at each developmental stage
7. Long stability of cDNA probes
8. Different loci may be identified by one probe
9. Heterologous genes may be used as probes
10. Any number of probes can be produced
11. Probes are producible for coding and silent sequences
12. Probes show the variability of flanking sequences
13. Several characters can be screened in the same sample

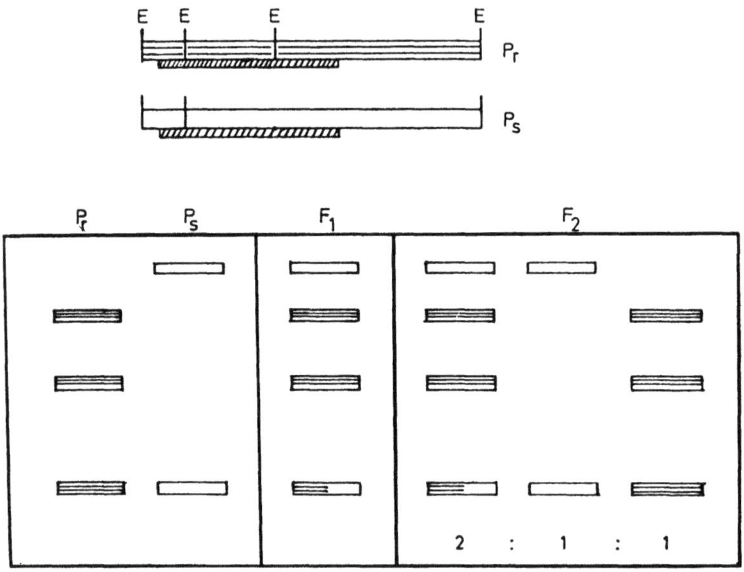

Fig. 57. Scheme of the inheritance of an RFLP. The *upper part* shows a segment from chromosomal DNA of a resistant (P_r) and a susceptible (P_s) parent. *E* indicates where a restriction enzyme occurs. Th e *hatched area* symbolizes the gene probe. The *lower part* demonstrates how the fragments of the upper DNA migrate in the electrical field. After hybridization with the probe, different patterns, the RFLPs, become visible. The RFLPs are shown for the parents, the F_1, and the segregating F_2

become possible to improve tomato varieties for this character by indirect selection for the linked RFLP marker.

Besides such predictions in breeding programs, RFLPs will be of importance in detecting variety characteristics and transferred genes. The role played by protein patterns for such diagnostic purposes will be overtaken by RFLPs. Furthermore, RFLPs can be used to explore the origin and evolution of crop plants, as was done for *Brassica* species and for potato. Figure 58 gives a diagram of the potato pedigree. By comparing primitive cultivars, wild species, and the characteristics of cultivated varieties, the location of economically important genes can be deduced. Figure 59 illustrates the maps obtained after probing different barley types. The particular probe used is able to differentiate winter and spring cultivars. The reliability of such tests is high when the molecular marker and the responsible gene are closely linked; otherwise, the correlation will be broken by a crossover event. Consequently, one needs a very dense RFLP map to make reliable judgements. If the map becomes very dense, it might be possible to find not only a molecular marker but the responsible gene itself. Thus, RFLP mapping is not only a procedure for genome analysis but also a method for physical gene diagnoses.

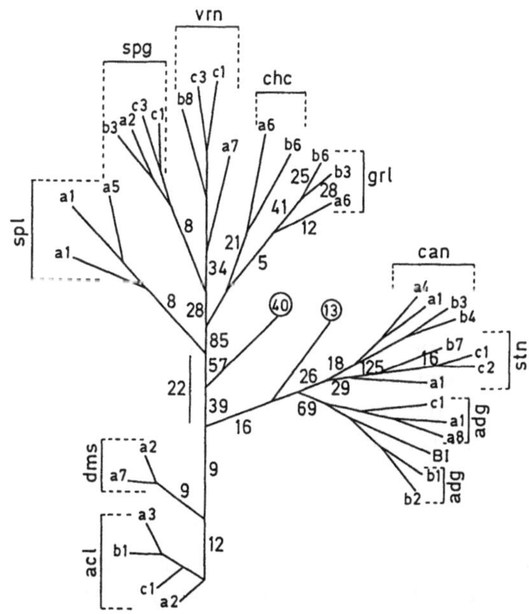

Fig. 58. Phylogenetic tree of *Solanum* species due to the genetic distance of RFLPs. *S. tuberosum* forms one group with ssp. *andigena*, *S. stenotomum*, and *S. canasense*. A second group is formed by *S. spars ipilum*, *S. berthaultii*, *S. kurtzianum*, *S. gourlayi*, *S. vernei*, *S. spegazzinii*, *S. chacoense*, and *S. megisacrolobum*. Two further branches are formed by *S. demissum* and *S. acaule* and by *S. polyadenium*, *S. pinnatisectum*, *S. stoloniferum*, and *S. etuberosum*. This is in good agreement with classical data (Debener et al. 1990)

Fig. 59. Phenotypic similarities documented by similar RFLP patterns. Barley DNA restricted with BamHI and hybridized after Southern blotting with hordein B probe pBSL 4

A physically identified gene, or DNA portions of the gene, is surely the most reliable probe in selecting for desired characters. The principle steps for gene identification were described in Section 3.6.3 under the aspect of gene transfer. It should be mentioned here, however, that it will often be easier to investigate a sexually produced population with gene probes for molecular markers or even directly for the desired traits. If a character is available in a heritable form, a simple and efficient selection procedure can be more economical than gene transfer. In addition, the problem as to how a foreign gene will behave in the genome of a given variety is not important in this approach. Furthermore, none of the problems related to releasing the transgenic plants into the open environment are significant.

For the identification of molecular markers, very low quantities of plant tissue are sufficient. The technique can be improved even further by using the polymerase chain reaction (PCR technique), which allows, e.g., in the case of barley, the correct hybridization of a DNA amount equivalent to less than 20 cells. The principle of the PCR technique is given in Fig. 60.

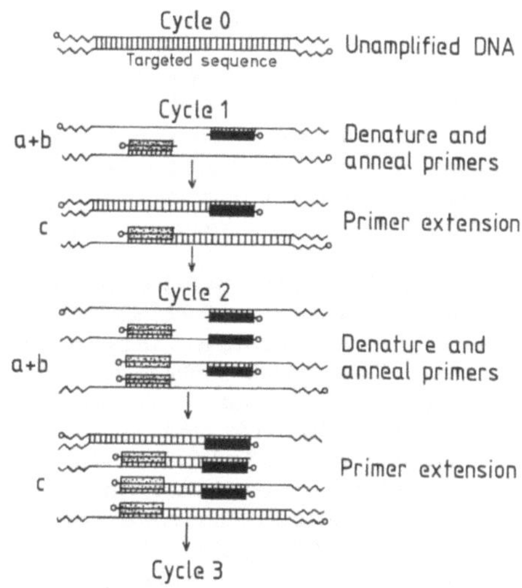

Fig. 60. The principle of the polymerase chain reaction

The use of molecular markers will become easier and more informative as more useful genes are known physically. It must be stressed, however, that these molecular techniques demand a very sophisticated knowledge of classical genetics. Molecular markers are only as good as the phenotypes that can be characterized. Classical genetics and molecular techniques are a unit, each one being the prerequisite for progress in the other field.

3.8 Cited and Recommended Literature for Chapter 3

Alfermann AW, Spieler H, Reinhard E (1985) Primary and secondary metabolism of plant cell cultures. In: Nemann KH et al. (eds.) Plant Cell Cultures. Springer Berlin 316–322

Arber W, Dussiox D (1962) Host specificity of DNA produced in *Escherichia coli*. Host controlled modification of bacteriophage lambda. J Mol Biol 5:18–24

Austin S, Baer MA, Helgeson JP (1985) Transfer or resistance to potato leaf roll virus from *Solanum brevidens* into *Solanum tuberosum* by somatic fusion. Plant Sci Lett 39:75–82

Bartelmess I (1964) Merkmalskorrelationen und Selektionen bei *Arabidopsis thaliana* (L) Heynh. Z Pflanzenzücht 52:273–332

Bauer R (1957) The induction of vegetative mutations in *Ribes nigrum*. Hereditas 43:320–337

Beckmann JS, Soller M (1986) Restriction fragment length polymorphisms and genetic improvement of agricultural species. Euphytica 35:111–124

Bergann F, Bergann L (1962) Über Umschichtungen (Translokationen) an den Sproβscheiteln periklinaler Chimären. Züchter 32:110–119

Beyer EY (1976) Silver ion: a potent antiethylene agent in cucumber and tomato. Hort Sci 11:195–196

Binding H, Nehls R, Schieder O, Sopory SK, Wenzel G (1978) Regeneration of mesophyll protoplasts isolated from dihaploid clones of *Solanum tuberosum* L. Physiol Plant 43:52–54

Blakeslee AF, Belling J (1924) Chromosomal mutations in the Jimson weed, *Datura stramonium*. J Hered 15:195–206

Broertjes C (1969) Mutation breeding of *Streptocarpus*. Euphytica 18:333–339

Buiatti M, Scala A, Bettini P, Nascari G, Morpurgo R, Bogani P, Pelligrini G, Gimelli F, Venturo R (1985) In vitro differential response of resistant and susceptible carnation cvs. to fungal and abiotic elicitors and to *Fusarium oxysporum* f.sp. *dianthi* culture filtrates. Theor Appl Genet 70:42–47

Chase SS (1974) Breeding diploid species. In: Kasha KJ (ed) Univ Guelph Ontario, pp 211–230

Coen DM, Bedbrook JR, Bogorad L, Rich A (1977) Maize chloroplast DNA encoding the large subunit of ribulose biphosphate carboxylase. Proc Natl Acad Sci USA 74:5487–5491

Debener T, Salamini F, Gebhardt C (1990) Phylogeny of wild and cultivated *Solanum* species based on nuclear restriction fragment length polymorphism (RFLP). Theor Appl Genet 79:360–368

Deimling S, Zitzlsperger J, Wenzel G (1988) Somatic fusion for breeding potatoes. Plant Breed 101:181–189

Den Nijs APM, Visser DL (1980) Induction of male flowering in gynoecious cucumbers (*Cucumis sativus* L.) by silver ions. Euphytica 29:273–280

Dixon LK, Leaver CJ, Bretell RIS, Gengenbach BG (1982) Mitochondrial sensitivity to *Drechslera maydis* T-toxin and the synthesis of a variant mitochondrial polypeptide in plants derived from mayze tissue culture with Texas cytoplasm. Theor Appl Genet 63:75–80

Dorè C (1974) Production de plantes homozygotes mâles et femelles à partir d'anthères d'Asperge. C R Acad Sci, Paris, Ser D278:2135–2138

Dorè C (1987) Application of tissue culture to vegetable crop improvement. In: Green et al. (eds) Plant tissue and cell culture. AR Liss, Inc New York, pp 419–432

Dressler O (1958) Zytologische Untersuchungen an diploidem und polyploidem Spinat (*Spinacia oleracea* L.) unter besonderer Berücksichtigung der Geschlechtsvererbung als Grundlage einer Inzucht-Heterosis-Züchtung. Z Pflanzenzücht 40:385–424

Dressler O (1973) Erfahrungen bei der Vermehrung und Züchtung monözischer Spinatsorten (*Spinacia oleracea* L.). Z Pflanzenzücht 70:108–128

Dudits D, Maroy E, Praznovszky T, Györgyey J (1986) Gene transfer between distantly related plants by fusion with irradiated protoplasts. In: Somers DA et al. (eds) Abstr IAPTC Congr, Univ Minnesota, Minneapolis, p 20

Dudley JW, Moll HR (1969) Interpretation and use of estimates of heritability and genetic variances in plant breeding. Crop Science 9:257–262

Ellison JH (1986) *Asparagus* breeding. In: Basset MJ (ed) Breeding vegetable crops. Avi, Westport, pp 521–569

Eresen H (1966) Die Verwendung von Künstlichen Artbastarden – insbesondere Additionsbastarden – in der Züchtung von *Tagetes* L. Z Pflanzenzücht 56:201–250

Foroughi-Wehr B, Wenzel G (1989) Androgenetic haploid production. IAPTC Newslett 58:11–18

Frandsen HN, Frandsen KJ (1948) Polycross-Methoden. Nord Jordbruks forsk 7–8:239–261

Franke G, Fuchs A (1980) Nutzpflanzen der Tropen und Subtropen, Band IV, Pflanzenzüchtung. Hirzel, Leipzig

Frankel R, Galun E (1977) Pollination mechanisms, reproduction and plant breeding. Monogr Theor Appl Genet 2. Springer, Berlin Heidelberg New York

Franken S (1979) Untersuchungen zur Vererbung vegetativer Merkmale und des Hermaphroditismus sowie zur Ertragsfähigkeit determinierter Einlegegurken (*Cucumis sativus* L.). Thesis, Univ Hannover

Friedt W, Foroughi-Wehr B (1983) Field performance of androgenic doubled haploid spring barley from F1 hybrids. Z Pflanzenzücht 90:177–184

Fromm MF, Taylor LP, Walbot V (1986) Stable transformation of maize after gene transfer by electroporation. Nature 319:791–793

Galun E (1961) Study of the inheritance of sex expression. The interaction of major genes with modifying genetic and non-genetic factors. Genetica 32:134–163

Galun E (1974) The use of genetic sex types for hybrid seed production in *cucumis*. In: Moav R (ed) Agricultural genetics-selected topics, Wiley & Sons, New York Toronto, pp 23–56

Gaul H (1961) Studies on diplontic selection after X-irradiation of seeds. In: IAEA (ed) Effects of ionizing radiation on seeds. Wien, pp 117–138

Gaul H (1966) Züchterische Bedeutung von Kleinmutationen. Z Pflanzenzücht 55:1–20

Gengenbach BC, Green GE, Donovan CM (1977) Inheritance of selected pathotoxin resistance in maize plants regenerated from cell cultures. Proc Natl Acad Sci USA 74:5113–5117

Gleba YY, Sytnik KM (1984) Protoplast fusion. Monogr Theor Appl Genet 8. Springer, Berlin Heidelberg New York

Glimelius K, Fahlesson J, Landgren M, Sjödin C, Sundberg E (1990) Improvement of the *Brassica* crops by transfer of genes from alien species via somatic hybridisation. In: Nijkamp et al. (eds) Progress in plant cell and molecular biology. Kluwer, Dordrecht, pp 299–304
Guha S, Maheshwari SC (1964) In vitro production of embryos from anthers of *Datura*. Nature 204:497
Hagberg A (1962) Production of duplications in barley breeding. Hereditas 48:243–246
Hagberg A, Hagberg G (1980) High frequency of spontaneous haploids in the progeny of an induced mutation in barley. Hereditas 93:341–343
Hauptmann RM, Ozias-Akins P, Vasil V, Tabaeizadeh Z, Rogers SG, Horsch RB, Vasil IK, Fraley RT (1988) Transient expression of electroporated DNA in mónocotyledonous and dicotyledonous species. Plant Cell Report 6:265–270
Helgeson JP, Kemp JD, Haberlach GT, Maxwell BP (1972) A tissue culture system for studying disease in tobacco callus cultures. Phytopathology 62:1439–1443
Hoekema A, Huisman MJ, Molendijk L, v.d. Elzen PJ, Cornelissen BJC (1989) The genetic engineering of two commercial potato cultivars for resistance to virus X. Bio/Technology 7:273–278
Horsch R, Fraley R, Rogers S, Fry J, Klee H, Shah D, McCormick S, Niedermeyer J, Hoffmann N (1987) *Agrobacterium*-mediated transformation of plants. In: Green Ce et al. (eds) Plant tissue and cell culture. AR Liss, New York, pp 317–329
Iwanaga M, Peloquin SJ (1982) Origin and evolution of cultivated tetraploid potatoes via 2n gametes. Theor Appl Genet 61:161–169
Janick J, Iizuka M (1962) Sex determination in spinach. In: Lechreiner A, Goeseels P (coord eds) 16th Int Hort Congr, Brussels II:82–88
Janick J, Mohaney DL, Pfahler PL (1959) The trisomics of *Spinacia oleracea*. J Hered 50:47–50
Jenkins C (1963) Secale additions and substitutions to common wheat. Hereditas Suppl 2:301–312
Jensen CJ (1976) Chromosome doubling techniques in haploids. In: Kasha KJ (ed) Haploids in higher plants, University of Guelph, Guelph pp 153–190
Johannsen W (1909) Elemente der exakten Erblichkeitslehre. Fischer, Jena
Kao KN, Michayluk MR (1974) A method for high frequency intergeneric fusion of plant protoplasts. Planta 115:355–367
Kao KN, Michayluk MR (1975) Nutritional requirements for growth of *Vicia hajastana* cells and protoplasts at very low population density in liquid media. Planta 126:105–110
Kasha KJ (ed) (1976) Haploids in higher plants. Univ of Guelph, Ontario
Kasha KJ, Kao KN (1970) High frequency haploid production in barley (*Hordeum vulgare*). Nature 225:874–875
Keller WA (1984) Anther culture of *Brassica*. In: Vasil IK (ed) Cell culture and somatic cell genetic of plants, vol 1. Acad Press, Orlando, pp 302–310
Keller WA, Melchers G (1973) The effect of high pH and calcium on tobacco leaf protoplasts fusion. Z Naturforsch 28c:737–741
Klee H, Horsch R, Rogers S (1987) *Agrobacterium*-mediated plant transformation and its further applications to plant biology. Annu Rev Plant Physiol 38:467–486
Klein TM, Gradziel T, Fromm ME, Sanford JC (1988) Factors influencing gene delivery into *Zea mays* cells by high velocity microprojectiles. Bio/Technology 6:559–563
Klopfer K (1965) Erfolgreiche experimentelle Entmischung und Umlagerung periklinaler Kartoffelklone. Züchter 35:201–214
Kobabe G (1972) Männliche Sterilität bei Spinat (*Spinacia oleracea* L.). Z Pflanzenzücht 57:114–122
Kubicki B (1982) Determination of sex and type of growth in spinach (*Spinacia oleracea* L.). Genet Polon 4:195–206
Kuhlmann U, Foroughi-Wehr B (1989) Production of doubled haploid lines in frequencies sufficient for barley breeding programs. Plant Cell Rep 8:78–81
Larkin PJ, Scowcroft WR (1981) Somaclonal variation -- a novel source of variability from cell cultures for plant improvement. Theor Appl Genet 60:197–214
Lelley T (1982) The inheritance in rye of crossability with common wheat. Tagungsbericht Nr.168, Akad Landw wissen DDR, Berlin, pp 191–198
Lichter R (1982) Induction of haploid plants from isolated pollen of *Brassica napus*. Z Pflanzenphysiol 105:427–434
Lind V (1983) A test system for the quantitative assessment of resistance to mildew in rye (Secale cereale L.) Phytopath Z 108:127–134

Linsmaier EM, Skoog F (1965) Organic growth factore requirements of tobacco tissue cultures. Physiol Plant 18:100–127

Löptien H (1979a) Untersuchungen zur Bestimmung der Geschlechtschromosomen beim Spinat (*Spinacia oleracea* L.). Z Pflanzenzücht 82:90–92

Löptien H (1979b) Identification of the sex chromosome pair in asparagus (*Asparagus officinalis* L.). Z Pflanzenzücht 82:162–173

Lörz H, Junker B, Schell J, De la Pena A (1987) Gene transfer in cereals. In: Green CE et al. (eds) Plant tissue and cell culture. AR Liss, New York, pp 303–316

Lower RL, Edwards MD (1986) Cucumber breeding. In: Basset MJ (ed) Breeding vegetable crops. Avi, Westport, pp 173–207

McClintock B (1950) The origin and behaviour of mutable loci in maize. Proc Natl Acad Sci USA 36:344–347

McCouch SR, Kochert G, Yu ZH, Wang GS, Khush GS, Coffman WR, Tanksley SD (1988) Molecular mapping of rice chromosomes. Theor Appl Genet 76:815–829

Melchers G (1946) Die Ursachen der besseren Anpassungsfähigkeit der Polyploiden. Z Naturforsch 1:160–165

Melchers G, Labib G (1970) Die Bedeutung haploider höherer Pflanzen für Pflanzenphysiologie und Pflanzenzüchtung. Ber Dtsch Bot Ges 83:129–150

Melchers G, Mohri Y, Watanabe K, Wakabayashi S, Harada K (1990) Method for the production of male-sterile tomato plants and male-sterile plants obtained thereby. Patent pending

Melchers G, Sacristan MD, Holder AA (1978) Somatic hybrid plants of potato and tomato regenerated from fused protoplasts. Carlsberg Res Commun 43:203–218

Morel G (1946) Essais de laboratoire sur le mildiou de la vigne: Rev Vitic 93:120–213

Murashige T, Skoog F (1962) A revised medium for rapid growth and bioassays with tobacco tissue culture. Physiol Plant 15:473–497

Nayar NM (1969) Considerations on overcoming intrasomatic selection during mutation breeding of vegetatively propagated plants. Theor Appl Genet 39:99–103

Nayar NM, Dayal TR (1970) A method of securing high frequencies of induced mutations in the potato. Z Pflanzenzücht 63:155–160

Nelson RS et al. (1988) Virus tolerance, plant growth, and field performance of tomato plants expressing coat protein from tobacco mosaic virus. Bio/Technology 6:403–407

Neuhaus G, Spangenberg G, Mittelstenscheid O, Schweiger HG (1987) Transgenic rapeseed plants obtained by the microinjection of DNA into microspore derived embryoids. Theor Appl Genet 75:30–36

Nijs AP den, Visser DL (1980) Induction of male flowering in gynoecious cucumbers (*Cucumis sativus* L.) by silver ions. Euphytica 29:273–280

Nitsch CN, Sheridan W, Neuffer G, Anderson S, Godard M (1982) The production of haploids from mayze and millet. In: Earle E, Demarly Y (eds) Variability in plants regenerated from tissue culture. Praeger, New York, pp 69–91

Novak JR (1973) A study of the effects of environmental and genetic factors on sex expression in cucumber (*Cucumis sativus* L.). Diss Abstr Intern B33, 2877-B-78B

Nybom N (1961) The use of induced mutations for the improvement of vegetatively propagated plants. Mutation and Plant Breeding, NAS-WRG 891:252–294

Osborn TC, Alexander DC, Fobes JF (1987) Identification of restriction fragment length polymorphisms linked to genes controlling soluble content in tomato fruit. Theor Appl Genet 73:350–356

Ou-Lee TM, Turgeon R, Wu R (1986) Expression of a foreign gene linked to either a plant-virus or a *Drosophila* promotor, after electroporation of protoplasts of rice, wheat and sorghum. Proc Natl Acad Sci USA 83:6815

Pelletier G, Primard C, Ferault M, Vedel F, Chetrit P, Renard M, Delourme R (1988) Use of protoplasts in plant breeding: cytoplasmic aspects. Plant Cell Tissue Organ Cult 12:173–180

Pike LM, Mulkey WA (1971) Use of hermaphrodite cucumber lines in development of gynoecious hybrids. Hort Sci 6:339–340

Pike LM, Peterson CE (1969) Gibberelin A4/A7 for induction of staminate flowers on the gynoecious cucumber (*Cucumis sativus* L.). Euphytica 18:106–109

Potrykus I, Paszkowski J, Saul MW, Negrutiu I, Shillito RD (1987) Direct gene transfer to plants: facts and future. In: Green CE et al. (eds) Plant tissue and cell culture. AR Liss, New York, pp 289–302

Potrykus I, Saul MW, Petruska J, Paszkowski J, Shillito R (1985) Direct gene transfer to cells of a graminaceous monocot. Mol Gen Genet 199:183–188
Raghavan V (1977) Applied aspects in embryo culture. In: Reinert B, Bajaj YPS (eds) Applied and fundamental aspects of plant cell, and organ culture. Springer, Berlin Heidelberg New York, pp 375–397
Ramage RT (1983) Heterosis and hybrid seed production in barley. In: Frankel R (ed) Heterosis. Monogr Theor Appl Genet 6. Springer, Berlin Heidelberg New York, pp 71–93
Ramage RT, Wiebe GA (1969) Use of chromosome aberrations in producing female parents of hybrids. In: IAEA (ed) Induced mutations in plants. Wien, pp 655–660
Redenbaugh KP, Viss P, Slade D, Fujii JA (1987) Scale-up: artificial seeds. In: Green Ce et al. (eds) Plant tissue and cell culture. AR Liss, New York, pp 743–493
Rieger R, Michaelis A, Green MM (1976) Glossary of genetics and cytogenetics, 4th edn. Springer, Berlin Heidelberg New York
Sacristan MD (1985) Selection for disease resistance in *Brassica* cultures. Heredity Suppl 3:57–63
Sanchez-Serrano JJ, Amati S, Keil M, Pena-Cortes H, Prat S, Willmitzer L (1990) The regulation of the potato proteinase inhibitor II gene expression. Plant Molecular Biology, VI NATO Inst Elmau, p 198
Schieder O, Vasil IK (1980) Protoplast fusion and somatic hybridization. Intern Rev Cytol Suppl 11B:21–46
Schmidt H (1970) Beiträge zur Züchtung apomiktischer Apfelunterlagen. III. Ihre Affinität mit Kultursorten in Abhängigkeit vom Virusgehalt. Z Pflanzenzücht 63:214–226
Scott JW, Baker RL (1976) Sex expression of single and 3-way cross cucumber hybrids with androecious pollinators. Hort Sci 11:243–245
Scowcroft WR, Larkin PJ (1988) Somaclonal variation, cell selection and genotype improvement. In: Fowler MW (ed) Comprehensive biotechnology, vol XX. Pergamon Press, Oxford, pp 153–168
Sears ER (1956) The transfer of leaf rust resistance from *Aegilops umbellulata* to wheat. Brookhaven Symp Biol 9:1–22
Sengbusch v R (1942) Süßlupine und Öllupine. Landw Jahrb 91:723–880
Shifris O, George WL, Quinones JA (1964) Gynodioecism in cucumbers. Genetics 49:285–291
Shillito RD, Carswell GK, Johnson CM, Dimaio JJ, Harms CT (1989) Regeneration of fertile plants from protoplasts of elite inbred maize. Bio/Technology 7:581–587
Sjödin C, Glimelius K (1989) Differences in response to the toxin sirodesmin PL produced by *Phoma lingam* (Tode ex Fr)Desm. on protoplasts, cell aggregates and intact plants of resistant and susceptible *Brassica* accessions. Theor Appl Genet 77:76–80
Skiebe K, Schmelzer K (1967) Zur Züchtung virusresistenter Hybridsorten beim Spinat (*Spinacia oleracea* L.). Z Pflanzenzücht 58:323–342
Smillie RM, Melchers G, v.Wettstein D (1979) Chilling resistance of somatic hybrids of tomato. Carlsberg Res Commun 44:127–132
Sneep J (1958) The breeding of hybrid varieties and the production of hybrid seed in spinach. Euphytica 7:119–122
Sneep J, Hendriksen AJT (eds) (1979) Plant breeding perspectives. Centre for Agricultural Publishing and Documentation, Wageningen
Soller M, Beckmann JS (1987) Cloning quantitative trait loci by insertional mutagenesis. Theor Appl Genet 74:39–378
Song KM, Osborn TC, Williams PH (1988) *Brassica* taxonomy based on nuclear restriction fragment length polymorphisms (RFLPs). 2. Preliminary analysis of subspecies within *B. rapa* (syn. *campestris*) and *B. oleracea*. Theor Appl Genet 76:593–600
Starlinger P (1980) A re-examination of McClintock's controlling elements in mayze in view of recent advances – in molecular biology. In: Leaver CJ (ed) Genome organization and expression in plants, Plenum Pub, New York, pp 537–551
Stegemann H, Schnick D (1985) Index of European potato varieties. Mitt Biol Bundesanstalt 227:3–128
Straub J (1950) Wege zur Polyploidie. Bornträger, Berlin
Sugiyama S, Suto T (1964) Studies on the artificial control of sex expresion in spinach. Bull Nat Inst Agr Sci Jpn, Ser D, 11
Tanksley SD, Hewitt J (1988) Use of molecular markers in breeding for soluble solids content in tomato – a re-examination. Theor Appl Genet 75:811–823

Tanksley SD, Young ND, Paterson AH, Bonierbale MW (1989) RFLP mapping in plant breeding: new tools for an old science. Bio/Technology 7:257

Tatlioglu T (1981) Stand und Möglichkeit der Züchtung von Einlegegurkensorten für eine maschinelle Ernte unter besonderer Berücksichtigung des determinierten Wuchstypes. Industr Obst-u Gemüseverwertg 66:303–311

Tatlioglu T (1983) Einfluß des Geschlechts des männlichen Hybridelters auf die Konstanz der Weiblichkeit der F1-Hybriden bei Einlegegurken (*Cucumis sativus* L.). Z Pflanzenzücht 91:140–153

Toriyama K, Hinata K (1988) Diploid somatic-hybrid plants between cultivars in rice. Theor Appl Genet 76:665–668

Tsuchiya T (1986) Gene analysis and linkage studies in barley. In: Yasuda S, Konishi T (eds) Barley Genetics V, Sanyo Press, Okayama, pp 175–187

Uchimiya H, Fushimi T, Hashimoto H, Harade H, Syono K, Sugaware Y (1986) Expression of a foreign gene in callus derived from DNA-treated protoplasts of rice (Oryza sativa L.) Mol Gen Genet 204:204–207

Uhrig H (1981) Regeneration of protoplasts of dihaploid potato plants bleached by a herbicide (San 6706). Mol Gen Genet 181:403–405

Uhrig H, Wenzel G (1987) Breeding for virus and nematode resistance in potato through microspore culture. In: Bajaj YPS (ed) Biotechnology in agriculture and forestry, vol 3. Springer, Berlin Heidelberg New York, pp 346–357

Vardi A, Epstein E, Breiman A (1986) Is the *Phytophthora citrophthora* culture filtrate a reliable tool for the in vitro selection of resistant *Citrus* variants? Theor Appl Genet 72:569–574

Vasil IK, Vasil V (1980) Clonal propagation. Intern Rev Cytol Suppl 11A:145–173

Virnich H (1967) Untersuchungen über das Verhalten der männlichen Sterilität und anderer Eigenschaften bei polyploiden Zwiebeln (*Allium cepa* L.) als Grundlage für eine Nutzung in der Hybridzüchtung. Z Pflanzenzücht 58:205–244

Webb RE, Thomas CE (1976) Development of F1-spinach hybrid. Hort Sci 11:546

Wenzel G, Brüning H, Dehmer K, Foroughi-Wehr B, Graner A, Möllers C, Schuchmann R (1990) Molecular biology and the development of improved diagnostic techniques. EAPR (in press)

Wenzel G, Brüning H, Foroughi-Wehr B, Jahoor A, Herrmann RG (1989) Identification of genes for disease resistance. Vorträge Pflanzenzüchtung 16:231–241

Wenzel G, Foroughi-Wehr B (1984) a: Anther culture of *Solanum tuberosum*. In: Vasil IK (ed) Cell culture and somatic cell genetics of plants, vol 1. Academic Press, Orlando, pp 293–301

Wenzel G, Foroughi-Wehr B (1984) b: Anther culture of cereals and grasses. In: Vasil IK (ed) Cell culture and somatic cell genetics of plants, vol 1. Academic Press, Orlando, pp 311–327

Wenzel G, Schieder O, Przewozny T, Sopory SK, Melchers E (1979) Comparison of single cell culture derived *Solanum tuberosum* L. plants and a model for their application in breeding programs. Theor Appl Genet 55:49–55

Werner D (1987) Pflanzliche und mikrobielle Symbiosen. Thieme Stuttgart

Westergaard M (1958) The mechanism of sex determination in dioecious flowering plants. Adv Genet 9:217–281

Wricke G (1968) Untersuchungen zur Vererbung des Geschlechts bei *Asparagus officinalis* L. I. Ein Majorfaktor für die Ausprägung des Andromonöziegrades. Z Pflanzenzücht 60:201–211

Wricke G (1973) Untersuchungen zur Vererbung des Geschlechts bei *Asparagus officinalis* L. II. Y-chromosomgebundene Unterschiede im Andromonöziegrad. Z Pflanzenzücht 70:91–98

Ye JM, Kao KN, Harvey BL, Rossnagel BG (1987) Screening salt tolerant barley genotypes via F1 anther culture in salt stress media. Theor Appl Genet 74:426–429

Zenk MH et al (1977) Formation of the indole alkaloids serpentine and ajmalicine cell suspension cultures of *Cataranthus roseus*. In: Barz W, Reinhard E, Zenk MH (eds) Plant tissue culture and its biotechnological application. Springer, Berlin Heidelberg New York, pp 27–43

Zenk MH, El-Shagi H, Schulte U (1977) Anthraquinone production by cell suspension cultures of *Morinda citrifolia*. Planta Medica Suppl 79–101

Zilm B (1966) Zytogenetische Untersuchungen an *Asparagus officnalis* L. in ihrer Bedeutung für die Züchtung rein männlicher und polyploider Sorten. Z Pflanzenzücht 56:1–26

4 Plant Breeding and Plant Production

World nutrition requirements for a continually growing population cannot exclusively be met by methods of plant breeding. In the production of food plants (Table 23 shows yields of the 22 most important food plants of the world for 1975 and 1987), the fields of plant production, plant physiology, and plant pathology, including agricultural chemistry, are of equally central significance. The yield increase in plant production achieved so far can be attributed about one-third each to progress in breeding, plant production, and agricultural chemistry. Because of the close interconnection of breeding with these related disciplines, this chapter will deal with their direct contribution to breeding. This presentation can only give a few details about neighboring fields; for more in-depth information, refer to special publications of plant production, physiology, and phytopathology.

4.1 Plant Breeding and Farming Systems

There are multiple relationships and interactions between the specialty fields of plant breeding and plant cultivation that make it necessary for the plant breeder to closely cooperate with the plant grower, in order to make his work as effective as possible. The diverse reaction of varieties to different factors in the environment, such as the respective yearly weather conditions and cultivation technique, is determined by the genotype of the variety. Therefore, for each new variety, the optimum conditions with regard to sowing time, sowing density, spacing, timing, and type and amount of mineral fertilizer used, as well its suitability for or adaptation to different cultivation regions, must be ascertained by trials performed over several years.

The optimum cultivation factors are determined, among other things, by the yield structure of a variety, the components of which are stand density (number of spike-carrying culms per plant), number of grains per ear, and single-kernel weight. In varieties with an extremely unilateral yield structure such as the one-stalked and long-eared wheat varieties of northern Italy, the stand density necessary for optimum yield can only be achieved by high sowing densities. Furthermore, strong nutrient applications must be administered, to enable the numerous flowers of the large ears to form grains with a high single-kernel weight by means of the assimilating organs of the plants.

Shortcomings of a variety in terms of its suitability for the cultivation methods used in practice, for utilization of the yield products, or concerning its insufficient adaptation sometimes only become apparent over the course of years,

Table 23. World plant production in 1975 and 1987 (according to FAO yearbooks)

Cultivated species	World harvest in million tons 1975	1987		World harvest in million tons 1975	1987
1. Sugarcane	650	968	12. Pulses	20	53
2. Wheat	360	517	13. Oats	50	47
3. Maize	300	457	14. Oranges	30	44
4. Rice	320	454	15. Bananas	35	39
5. Sugar beets	261	297	16. Cabbages	15	38
6. Potatoes	300	285	17. Rye	30	32
7. Barley	170	179	18. Onions	15	25
8. Soybeans	60	98	19. Rapeseed	8	23
9. Grapes	60	65	20. Groundnuts	20	20
10. Tomatoes	35	61	21. Sunflower	10	20
11. Sorghum	95	59	22. Cotton (fiber)	25	17

with its wider distribution in cultivation under varying weather conditions. Under the optimal trial conditions of a breeding or other experimental station, shortcomings of a variety are often concealed. Due to his experience with existing varieties and furthermore his knowledge of the future requirements that a modified growing technique will impose on the new varieties, the plant grower is particularly competent in setting special breeding aims for the plant breeder.

Recent production aims and production methods regarding efficiency of working methods, above all the saving of labor, determine the requirements of performance, quality, and structure of varieties. Thus, mechanization by combine harvesting requires, among other things, varieties particularly resistant to lodging with a narrow grain/straw ratio, which do not shatter easily, thus avoiding larger losses due to overripeness. Furthermore, modern cereal cultivation requires varieties that do not react unfavorably to the application of herbicides, fungicides, insecticides, and growth regulators. The reaction of wheat varieties to chlorocholine chloride (CCC) has become a new important breeding aim. CCC is a growth regulator, for the development of a shorter and more stable stalk, and increases yield because higher nitrogen applications are then possible. In sugar beet, the need for monogermic seeds to reduce hand labor is met by monocarpic varieties.

As the history of plant breeding teaches us, the aims set in plant breeding are not always determined by the requirements of plant cultivation alone, but at times also by the breeder himself. The breeder's own aims are based on his knowledge of the multiplicity of types of the cultivated plants and their related primitive types and wild species, of the history of the origin and rise of the cultivated plants and, furthermore, of the accumulated experience on the degree of combinations that can be achieved. The unexpected genotypes found by chance during the course of a certain breeding program, linked with the idea of their practical use, may occasionally also determine tasks and trends in plant breeding. Therefore, plant breeding can have a stimulating and modifying effect on plant production.

In this context, the search and discovery of rape mutants free of eruca acid is brought to mind, with which the rape oil is qualitatively improved, so that it is equivalent to oils of sunflowers, soy beans, and peanuts and is therefore also used in the margarine industry. An expansion of rape cultivation caused by increased demand is also favorable in terms of plant production, especially for crop rotation.

The development of polyploid varieties, based on initiatives taken by the breeders, is also mentioned in this context. These sometimes differ so much from diploid original varieties in their requirements and in the utilization of the yield products, that their cultivation technique and usage must be extensively modified. Finally, a successful development of new cultivated plants from wild types, e.g., from *Cuphea* species (Lythraceae), the seeds of which supply the medium-chained fatty acids desired by the petrochemical industry, can also give new impulses to the field of plant production.

An important task of plant production is to arrange environmental conditions in such a way that optimal yield per area and per time unit can be achieved, and at the same time a certain yield reliability can be expected. In order to reach these goals, the plant grower must make use of corresponding research results from plant physiology and plant pathology or must continue the development of their methods, so that they can profitably be applied in plant production.

No less than the plant grower, the plant breeder is also interested in plant physiology and plant pathology. The processes in metabolism and the development of plants are controlled by genes; the different behavior against pathogens is also genetically conditioned. Applying methods that were developed on the basis of plant physiological and plant pathological knowledge, the breeder is able to test the different behavior of genotypes of his breeding material and thus carry out his breeding program more effectively and with a higher prospect of success. For this reason the two following sections show the relationship of plant breeding to plant physiology and plant pathology by citing a few examples.

4.2 Plant Breeding and Plant Physiology

The purpose of plant breeding is to improve our cultivated plants from different aspects such as yield capacity, disease resistance, drought tolerance, harvest product quality, etc. The change in genetic constitution of the plant material, which is the result of the breeding work, is in almost all cases accompanied by a change in the physiological properties of the plants and frequently also by changes in their morphology, anatomy, or biochemistry. It is, therefore, of great importance for the plant breeder to become familiar with the physiological, morphological, and other characteristics of his plant material in order to achieve maximum efficiency in his breeding work. This statement is true for the application of "classical" breeding methods and strategies such as selection in populations, hybridization, polyploidy, and use of mutagens, but it is of particular importance when the new advances in molecular biology and cell and tissue culture are to be used in practical breeding work. We now have access to a number of more or less sophisticated methods for isolating single or complex genes, for transferring them, and for introducing them into selected plants or plant organs. However, these techniques are only of limited value, unless we more know precisely what particular physiological and related processes we want to affect, how these processes behave and how they are regulated and controlled. A short survey follows, which will present the major groups of physiological processes that are of special interest in this regard. Inevitably there will be omissions of certain problems or applications that may be of special interest to some readers. In such cases, their attention is directed to more comprehensive textbooks and reviews and to original articles written by specialists in the field.

4.2.1 Biomass Production

The primary goal of agriculture is to raise crops that produce large quantities of biomass per growing area and that apportion a maximum of this biomass to harvestable plant parts. A secondary, but in many cases equally important goal is to achieve the best possible quality of the harvested plant products. Also in silviculture biomass production is of utmost importance, whereas in horticulture the quality of the harvested plants or plant parts frequently is of greater significance than the production of large quantities of dry matter. The task of the plant breeder is to provide the grower with suitable plant material which, under the given environmental conditions, best fulfills these different goals.

A crucial question is whether or not increased total biomass production or optimized partitioning of the produced biomass should be given highest priority as a selection goal. For forage crops such as grasses and clovers, the answer is simple; here, increased total biomass is the most important breeding objective, since virtually all above-ground biomass is harvested and processed. Of course, this does not exclude the fact that quality factors like protein content are also of great importance, but they are secondary in the present context.

In cereals as in many other crops, the trend in breeding new cultivars during the last century has clearly been towards an increase in the proportion of harvestable products (= an increased harvest index). In fact, the total biomass production of modern cultivars of wheat and barley has been shown to be more or less the same as that in plant material grown around the year 1900. As a consequence of this trend, many crop physiologists have advocated that an increased harvest index should be given very high priority as a breeding objective for increased yield. A practical demonstration of the feasibility of such a strategy is the successful breeding of high-yielding cultivars of rice at the International Rice Research Institute (IRRI). Another promising approach is to screen for both harvest index and a second trait, e.g., growth rate, and to try to achieve a good combination of both in a single genotype. However, harvest index cannot be raised indefinitely and, according to several authors, we are now near the limit for that increase and will soon have reached the point where an increased total biomass production together with a maintained high harvest index is necessary for further increases in grain yield.

As already indicated, biomass production is the result of a number of interacting biological processes, which are controlled both by the genetic constitution of the plant material and by the environmental influence of the particular growing site. Among the biological processes that together determine the level of biomass production, photosynthesis is undoubtedly the most important one under nonstress conditions. The products of photosynthetic activity are either used for further plant growth, are stored as surplus material in different plant parts, or are consumed in respiratory processes (both in light and in darkness). A decrease in dry matter losses due to respiration is, therefore, an obvious breeding objective.

Under conditions of environmental stress (which in practice nearly always is present to a greater or lesser degree), processes such as uptake and use of water and nutrients may take over as major yield-limiting factors. Increased tolerance or resistance against such stresses appear as very important breeding goals under these circumstances. Increased resistance against disease and parasite stress may become crucial for similar reasons.

4.2.2 Photosynthesis

Photosynthesis, the process by which light energy coming from the sun is intercepted by the plant canopy and transformed into chemical energy needed for further biosynthetic activity, comprises an extremely complex series of biochemical, physiological, anatomical, and morphological features, which have to be precisely matched and timed if biomass production of the crop is to increase and become consistent. In order to compensate for temporary declines in the performance of the photosynthetic apparatus, plants have developed various strategies that enable them to counteract the negative effects of a reduced flow of photoassimilates. It is not surprising, then, that most attempts by plant breeders to select for an increased specific rate of photosynthesis have so far failed to result in increased biomass production. This does not mean, however, that plant breeding for increased photosynthetic production is doomed in the future; the problem instead is to find the right characters to select for under the prevailing conditions of cultivation.

Photosynthetic production depends on three basic features: (1) the size and spatial orientation of the light-intercepting green area (leaves as well as other green plant parts), (2) the duration of this area in an active state, and (3) the specific rate of photosynthesis per unit green area. Of these features, the first two exhibit the greatest genetic variation, and they have also been most successfully exploited in practical breeding. The selection for erect, slowly senescing leaves in rice is a classical example of the progress that can be made with such an approach. Another illustration of the potential of photosynthetic production is given by a change in leaf characteristics in sugar beet. In this crop it has been found that certain cultivars expand their leaf area twice as fast as others at low temperatures (7 °C). They are, therefore, able to attain an optimum leaf area index (LAI) much earlier during the growth season than the more slow-growing genotypes, and this increases quite substantially their capacity to produce dry matter. There are good reasons to believe that further improvements of the structure and duration of the light-intercepting canopy are still possible in many crops, particularly if the photosynthetic production during critical stages in plant development is considered.

It has become increasingly evident that in order to obtain high yields, an ample supply of photosynthetic products is essential, not only during the filling stage of storage organs like grains, fruits, roots, or tubers but also, and often even more important, during the phases of initiation and early growth of these organs. In cereals, for example, grain number is generally a more important yield determinant than grain weight, and the formation of either a large number of ears or a large number of grains per ear (or both) in most cases turns out to be a more important breeding objective than the attainment of large individual grains. Since the ultimate number of grains per unit of growing area is dependent on the extent to which the initially high numbers of primordia for ears, spikelets, and florets are reduced during ontogenesis, a sufficient supply of photoassimilates during these crucial stages is particularly important.

The partitioning of available photoassimilates among different plant organs is obviously under some kind of internal control. The exact nature of this control is not known, but it appears to be related to the growth activities of the different sink regions within the plant, and may be mediated with the help of appropriate plant hormones. At the moment, however, there are no suitable methods for a direct exploitation of these potential control mechanisms in practical breeding work.

The specific rate of photosynthesis varies considerably between and within species. It has therefore been an obvious challenge to find particularly efficient genotypes but, as already mentioned, the selected types have seldom exhibited an increased yield or even an increased biomass production. In contrast, it has frequently been observed that genotypes with high specific photosynthetic rates are inferior, with respect to yielding ability, to types with lower specific rates, but a better-adapted morphology or duration of the light-harvesting system. Thus, the prospects for increasing yielding ability by selecting for high specific rates of photosynthesis do not seem to be very bright at the moment. This situation may change, of course, if molecular biologists should dramatically succeed in improving the capacity of the CO_2-fixing enzymes (Rubisco primarily). Even then, however, there still remains the possibility that internal rate control will dominate over potential capacity.

4.2.3 Respiration

The respiratory processes in plants are always accompanied by a certain loss of dry matter, mainly carbohydrates, and are therefore often considered a priori as negative phenomena in the production process. This view is too simplified, however, since respiration supplies plants with both essential energy and with intermediate products for further metabolic activity. Respiration is thus of vital importance, both for the maintenance of existing structures and for continued growth. On the other hand, this does not exclude the fact that in certain cases reductions of nonvital respiration can be achieved and can even prove to be of practical value.

4.2.3.1 Photorespiration

The key enzyme in photosynthesis, Rubisco, has the unique double function of being both a carboxylase and an oxygenase. When acting as an oxygenase, Rubisco catalyzes the light-dependent uptake of O_2 and the formation of CO_2 in a complicated process designated photorespiration, which takes place concomitantly in three cell organelles, the chloroplasts, the peroxisomes, and the mitochondria. In the course of photorespiration, both newly synthesized carbon compounds and energy are lost. In C_4 plants, photorespiration is suppressed to a very large extent due to the presence of a very efficient CO_2-concentrating mechanism.

In C_3 plants, about 25% of the net carbon uptake is re-evolved immediately in photorespiration. It is, therefore, tempting to try to avoid these losses by searching for genotypes with reduced or eliminated photorespiration. Genotypes with reduced photorespiration have also been found, e.g., in ryegrass and *Panicum miliolides*, but it is still a matter of debate as to whether photorespiration is indispensable for C_3 plants or not. Several findings suggest that it acts as a "safety valve," which protects the plants against the accumulation of surplus radiant energy that may occur under high light or unfavorable temperature conditions, and which destroys the light-harvesting pigments in the chloroplasts. Until this question is answered it appears to be premature to begin plant breeding projects that aim at eliminating photorespiration.

4.2.3.2 Dark Respiration

Respiration in darkness exhibits two principal components related to the maintenance of existing structures and functions and to the synthesis of new tissues. Of these components, maintenance respiration accounts for a considerable proportion of the carbon loss in mature plants, and it has been suggested that biomass production in crop plants might be increased by selecting for genotypes with reduced maintenance respiration. Screening populations of tall fescue and perennial ryegrass for low respiration rate resulted in lines that out-yielded those lines with high respiration rates, as well as the parents of both. The higher crop growth rates of the low-respiration lines were also maintained during regrowth in the following generation, and the improvement in annual dry matter production in field trials was found to be about 8%. It is still an open question as to whether plants must pay some penalty for having low respiration rates; if not, selection should have favoured those genotypes already long ago.

4.2.4 Uptake and Use of Nutrients from the Soil

In addition to a steady flow of carbon fixed from the atmosphere, plants need to be supplied with sufficient amounts of nutrient elements from the soil. The ability to acquire and use these nutrients is under genetic control, hence the prospects of accomplishing successful plant breeding seem to be good. Whereas the possibilities of increasing (in cases of deficiency) or decreasing (in cases of toxicity) the uptake and further assimilation of different micronutrients were recognized and exploited long ago, the question of improving the acquisition potential for macronutrients (N, P, and K) has raised relatively little interest until recently. The reason for this lack of interest is due to the fact that it has been easier and more economical in the past to use fertilizers as a means of correcting macronutrient deficiencies. This situation has changed considerably in recent times, however, both due to rising costs for fertilizers and because of the increasing problems with leakage of nutrients and pollution of the environment. There is, thus, an increasing demand for crop cultivars that can trap as much as possible of the available macronutrients in the soil and/or that can use the absorbed nutrients in the most efficient way for the production of biomass in desired sites within the plants. The opportunities to breed for special "low-input" cultivars have also been much discussed during the last decade.

When considering the possibility of selecting for macronutrient-efficient genotypes, one must distinguish between at least three fundamentally different problems that affect the uptake, translocation, and utilization of the nutrients. This creates a rather complex hereditary situation, since each of the three above-mentioned alternatives is complex in itself, and implies the interaction of a number of more or less independent genes. This is in strong contrast to that which applies for the uptake and use of micronutrients, a process that is frequently dependent mainly on the action of a single gene and, consequently, can be handled much more easily in breeding work.

An obvious approach to breeding for increased nutrient uptake capacity is to select for an improved root system. Root characteristics of different kinds are heritable but are also much influenced by the environment. Plant roots possess a great potential for adapting to different soil conditions. It is therefore difficult, for instance, to gather root growth data that are consistent and can be used in practical breeding work. Nevertheless, root measurements may be of value under certain circumstances, e.g., in connection with studies on drought resistance, and will be further discussed in the section dealing with water stress.

Nutrient uptake is dependent not only on the size and distribution of the root system, but also on its physiological activity. Nutrients may enter the roots passively with the transpiration stream, but their influx and efflux is primarily regulated by the physiological activity of the root and shoot system. A number of kinetic parameters (V_{max}, K_m, C_{min}) are frequently used to characterize the active component of nutrient uptake into the roots, and it has been established that different genotypes show considerable variation with respect to the magnitude of these parameters when determined under standard conditions of cultivation. However, the response is highly dependent on external factors such as nutrient level and root temperature, as well as on plant age and shoot metabolism. It is, therefore, dangerous to use results from isolated kinetic measurements as selection criteria. Another obstacle is that such measurements are normally rather time-consuming and expensive.

After entering into the plant through the root system, nutrients must be translocated into other plant parts and converted in different ways, either prior to their movement or after their arrival at the intermediate or final destination. The translocation and conversion processes are subject to genotypic control as well, again in conjunction with influences from the environment and the ontogenetic status of the plants. The combined characteristics of uptake, translocation, and conversion determine how much of a nutrient is accumulated within a plant or a particular plant organ. This efficiency of nutrient acquisition (defined as nutrient uptake efficiency and measured as the ratio between accumulated and available nutrients) may be of considerable interest to the plant breeder if he wants to select for genotypes that are effective in accumulating nutrients which are present in abundant supply. Of still greater interest, however, is the case where an improved utilization of the accumulated nutrients is desired. Nutrient utilization efficiency can be defined as the ratio between yield and amount of accumulated nutrient (both per plant or for some other unit). A high nutrient utilization efficiency may be of particular importance under conditions of poor nutrient supply. As an example of successful breeding for improved nutrient utilization efficiency, the screening for phosphorus-efficient rice genotypes at IRRI might be mentioned.

With respect to the relative importance of different nutrient elements, nitrogen holds a prominent position, mainly because of its direct connection to protein synthesis and, thereby, to growth and development of whole plants and their different organs. Nitrogen metabolism is, moreover, closely related to carbon metabolism and is thus of the utmost importance for biomass production and yield formation. The partitioning of carbon assimilates formed in various green tissues (sources) to other plant parts (sinks) represents one of the basic features of yield formation. Both sink capacity and sink activity are strongly related to the presence of sufficient amounts of nitrogen to sustain vigorous growth and metabolism of the different sink regions. In cereals, much emphasis has been laid in the past upon factors influencing grain growth after anthesis. However, there has recently been a shift towards an increasing recognition of the importance of events prior to anthesis, i.e., when the sink capacity of ears and ear structures is generated. Nitrogen plays a central role in this process and an important breeding objective is to select for genotypes that can most efficiently use the available nitrogen resources for this purpose.

4.2.5 Nitrogen Fixation

A special case of nitrogen acquisition can be found in those plant species that are able to fix atmospheric nitrogen, particularly the legumes among higher plants. The ability of these plants to fix nitrogen is due to the existence of a highly specialized symbiosis between the plants and certain *Rhizobium* bacteria enclosed in special nodules on the roots. Attempts to increase the nitrogen fixation efficiency of the legume-*Rhizobium* symbiosis by breeding may be directed towards three main areas: (1) the nitrogen-fixing capacity of the bacteria, (2) the interaction between the two partners of the symbiosis, and (3) the efficiency of the host plant in producing and supplying assimilates rich in energy, mineral elements, and other components essential for the function of the bacteria.

The nitrogen-fixing bacteria are relatively strongly host-specific, but there is great variation among the various *Rhizobium* strains with respect to infection ability and fixation efficiency. The use of particularly efficient *Rhizobium* strains may be difficult in practice, however, since it has been found in many cases that endogenous, less efficient populations of *Rhizobium* bacteria in the soil compete successfully with externally applied, efficient strains, thereby preventing the establishment of superior combinations of efficient bacteria and selected host plants.

Interactions between bacteria and hosts may occur at many levels, starting with the recognition, infection, and nodulation, continuing with the mutual exchange of assimilates and other substances between the partners, and ending with the inactivation or death of the nodules. These different processes are controlled by genes located either on the bacteria or on the hosts or on both and there are thus, at least in theory, ample possibilities of improving the efficiency of the symbiosis by selecting for single or combined N_2-fixation characteristics.

In the first days after germination the cotyledons of the seeds furnish the necessary nitrogen assimilates, but when this source is exhausted, the plants tend to be nitrogen-starved until their nitrogen-fixing system is fully operative. In many legumes, externally applied nitrogen fertilizers antagonize the formation and function of bacteria nodules, which further delays the attainment of vigorous plant growth. It would be of great advantage for the cultivation of these species if genotypes could be selected that can tolerate a moderate "starting dose" of nitrogen fertilizer better than existing ones or even benefit from it.

Since nitrogen fixation is a highly energy-demanding process, the supply of photosynthates to the bacteria nodules is very critical for their proper functioning. An increased photosynthetic capacity of the host plant is, therefore, an obvious breeding objective, with particular emphasis laid on the maintenance of sufficient photosynthetic activity also during later stages in plant development. When the pods and particularly the seeds start to grow after flowering, this introduces an extensive shift in the whole pattern of assimilate translocation and distribution. Pods and seeds represent extremely strong sinks for assimilates of different kinds and compete successfully with other plant parts, especially the roots, for the available resources. It is, therefore, very important that enough photosynthates are produced and translocated to the roots to ensure the further activity of the nitrogen-fixing bacteria. If not, transport of nitrogen assimilates from the roots will cease very quickly, and breakdown of leaf proteins will start fairly soon to compensate for the loss of root assimilates. The degradation of leaf proteins results in a rapid decline in photosynthetic production with subsequent retardation of seed growth. Plant breeding efforts should, therefore, aim at the selection or construction of genotypes that are able to sustain a reasonably extensive transport of photosynthates to the roots also during the period after flowering. However, the problem of source-sink relationships is particularly complicated in the legumes, and several other aspects besides those mentioned above (also to include questions of crop management) must be considered.

A great deal of speculation in the past has been about the possibility of transferring the nitrogen-fixing system, and specifically the nif-genes, into non-legume plants, thereby enabling these plants to be self-supporting with nitrogen. The obstacles to such a transformation are formidable, however, and although it is, e.g., possible to establish new *Rhizobium*-host plant combinations, it is hard to envisage how the whole spectrum of difficulties can be overcome in the foreseeable future.

4.2.6 Water Stress

Most parts of the world are subject at times to more or less severe drought, and lack of available water is, in fact, one of the most important factors in reducing crop yields. However, in some places the opposite is true; in these areas, excess water due to heavy rainfall or other reasons may impose great damage to crops. By applying agronomic practices such as irrigation or drainage, farmers have tried to counteract these stresses as much as possible. Yet there are limits to how much can be achieved in this way and other possibilities must be considered as well. Foremost among these is breeding for stress-tolerant cultivars of the crops grown in these regions.

Three major types of environment are associated with drought stress, and breeding strategies must be adapted to the particular circumstances. In the first type of environment, plants must rely completely on moisture accumulated in the soil during a previous wet period. The beginning, duration, and degree of a developing stress is known with reasonable precision, hence breeding efforts can be based on clear-cut premises. The second type of moisture environment is one in which the crop is grown during that part of the year when rainfall is expected to occur. The amount and periodicity of precipitation are rather erratic, however, which creates different drought stress situations during different years and makes breeding work more difficult. The third type of moisture environment is characterized by a sufficient amount of available water during most of the growing season, but with spells of drought during some periods and some years. Since the occurrence of these spells is irregular, breeding for resistance against this type of stress is also rather difficult.

Breeding for drought resistance can be based on one of four fundamental approaches. The first is to breed for high yields under conditions of unlimited water supply and to assume that the selected genotypes will also be superior under regimes of water stress. The second approach is also based on selection for high yields, but now under the same stress situation as the crop will later be subjected to. The third and fourth approaches, finally, include the incorporation of specific morphological and physiological traits, believed to increase drought resistance, into plant material of conventional breeding programs. Whereas a combination of various morphological and physiological selection criteria is used in the third approach, not more than one single drought resistance character is taken into account in the fourth one. It is impossible to make a general judgement as to which approach is the best one from a practical point of view; all have their merits and drawbacks, but obviously the choice depends to a large extent on the type of environment for which the crop is intended.

The water status of a growing plant is basically controlled at two levels: by the uptake of water through the root system (and its further transport to other parts of the plant) and by the loss of water via the stomata and the cuticle. Breeding strategies may, therefore, be focused on one of these areas with the aim of either increasing water uptake or reducing water loss. These strategies have the common objective of avoiding or delaying the occurrence of stress as much as possible, whereas a third strategy has the aim of increasing the tolerance of the plants against the developing stress. The basis for this last-mentioned approach is the fact that many crop plants accumulate solutes as a response to stress, and thereby exhibit an osmotic adjustment that can help the plants to survive and even continue to grow at a certain water deficit.

An obvious approach to increase the water uptake capacity of crop plants is to select for a root system that penetrates the soil deeper and faster, thereby enabling the plants to utilize water resources that would have otherwise been inaccessible. In cereal plants the seminal roots are especially suited for this purpose, and it has been shown that there is considerable genetic variation in the growth and extension of the root system, which can be exploited in breeding work. From a practical point of view, a major difficulty is that root measurements on adult plants in the field are extremely laborious, hence expensive, and are therefore of only limited value as selection criteria. Attempts have been made, however, to correlate root (and shoot) measurements on oat seedling plants cultivated in nutrient solutions, with the yield response of the same material grown under simulated drought conditions (achieved with the aid of movable rain shelters and a drip irrigation system). A strong correlation was found between seminal root length (or a calculated drought resistance index), on the one hand, and field drought susceptibility index, on the other, and it was suggested that these seedling characteristics might be used as selection criteria in the breeding of oats for specific drought-endangered areas in Sweden. For the investigation of more limited materials, special root inspection devices (perspex tubes, boxes with glass windows, etc.) have proven to be useful in many cases.

Attempts to select for plant characteristics such as reduced stomata number, which are thought to decrease water loss in above-ground plant parts, have met with varying success. This is because reduced gas exchange through the stomata also impairs CO_2-uptake and, hence, decreases photosynthesis and dry matter production. A more promising approach is to select for genotypes which, by osmotic adjustment, are able to maintain their turgor potential under stress conditions and thus can sustain stomata opening, photosynthesis, and growth. The transpirational water loss of cut and wilting seedlings can easily be measured, and their leaf areas and other relevant parameters can also be calculated. Positive results have been obtained with this method.

During the last decade new methods have been developed for surveying disease or water stress effects on large numbers of plots in field trials. These are based on remote sensing via infrared aerial photography or IR-thermography, and also appear to have potential in plant breeding.

4.2.7 Temperature Stress

All plant species attain optimum growth and development at a temperature regime that is characteristic for that species. If the environmental temperatures deviate substantially from these optimum levels, the plants are subjected to a more or less pronounced stress. One of the main objectives of plant breeding is to produce cultivars that are adapted as closely as possible to the temperature regime that normally prevails at the intended place of cultivation. In addition, the plants must be able to tolerate or at least survive occasional more or less strong deviations from the normal temperature regime.

Temperature stress can occur at both low and high temperatures, the former being more common in temperate climates and the latter in subtropic and tropic regions. Many plant species have acquired the ability to harden, i.e., to adapt themselves to a gradually developing stress. This hardening ability is frequently a prerequisite of successful breeding for stress resistance.

Cold hardiness, i.e., the ability to withstand freezing temperatures, is an essential character for winter crops of the temperate zones. The testing of breeding material for cold hardiness is a difficult task, however. Although they provide the ultimate answer, field tests are often unreliable because of the fluctuating weather conditions within and between years. Artificial freeze tests require some relatively expensive equipment and, moreover, must be very carefully conducted if reliable results are to be obtained. Nevertheless, valuable results for breeding work can be obtained with some of these methods. Still more promising are methods based on measurements of electrical conductivity or impedance in hardened leaves, which would allow the screening (and further cultivation) of single plants on a large scale.

Some categories of plants, particularly those of tropical or subtropical origin, are often sensitive to temperatures between 0 and 10 °C. This chill sensitivity varies between genotypes of the same species and may, therefore, be improved by breeding efforts. Besides the more traditional methods (crossing with wild species, etc.) of introducing new gene material into the cultivated types, the use of mutagens in combination with cell culture techniques may be of value for this purpose.

Heat tolerance is another important aspect of temperature stress. In the field, water stress and temperature stress are often linked and it may, therefore, be difficult to distinguish between these two factors in ordinary yield trials. They can be easily separated in the laboratory or in a growth chamber, however, and it is clear that there are considerable differences with respect to heat tolerance between different genotypes. Such differences can also be exploited in practical breeding work, e.g., in the adaptation of potatoes from temperate or highland regions to subtropical or tropical conditions.

Preharvest sprouting in cereals is caused by a different type of combined water and temperature stress. In this case high moisture conditions, in combination with an inducive temperature regime during a period shortly before and after kernel ripening, may result in severe damage of the crop because of premature germination of the kernels. All cereals are to a greater or lesser degree susceptible to preharvest sprouting, and there are great genetic differences with respect to sprouting tendency within each species. Sprouting resistance is dependent on several physiological and biochemical characters (degree of dormancy, enzyme activity, etc.) and is in most cases controlled by more than one gene. The prospects for successful breeding are good and considerable progress has been made in recent years with respect to both theory and practical applications.

4.2.8 Salinity Stress and pH Stress

Two other stress situations that are very common in soils of different parts of the world and that deserve attention are caused by increased salt concentrations and/or extreme pH values. These types of stress are often exacerbated by human activities, and create increasing problems for agriculture in many regions. Both wild and cultivated plants display various degrees of adaptation to these stresses, and plant breeding has been successful in producing new cultivars with a considerable degree of resistance to high salinity or low pH in the soil. The selection work is rather straight-forward and many gene sources exist for increased resistance, both in wild and cultivated plant material.

4.2.9 Photoperiodic and Vernalization Requirement

The transition from a vegetative to a generative stage represents the most thorough and far-reaching change in the physiology of a plant. It also has great implications from an agronomic point of view, since the yielding properties of the various crop plants are determined to a large extent by the course of the different events that initiate and accompany this radical shift. The consecutive steps in the transition process are genetically controlled, but they are also strongly influenced by different environmental factors, particularly photoperiodism and temperature. There is great variability in the response of different genotypes to the critical environmental factors and, consequently, ample scope for plant breeding progress.

With respect to photoperiodic requirement, plants can be characterized as long-day types (at least 14 h light/day), short-day types (10-14 h light/day), or day-neutral types (10–24 h light/day). Photoperiod not only controls the onset of flowering, it may, for example, initiate the formation of tubers in potatoes and similar crops, and it also affects the time of maturity. Besides having a qualitative effect (flowering versus nonflowering, etc.), photoperiod also induces quantitative reactions. The photoperiodic effect is modulated both by other environmental factors, particularly temperature, and by the genetic constitution of the plants. It is the interaction among these different factors which, to a large degree, determines the yield potential of a specific cultivar.

Winter forms of the cultivated plants have a more or less pronounced vernalization requirement, i.e., they must be subjected to relatively low temperatures (near 0 °C) in order to shoot and to produce flowers. A high vernalization requirement is normally accompanied by a high degree of frost resistance and vice versa. Consideration must be given in breeding work to the vernalization requirement of the different crops at their particular growing localities. Artificial vernalization treatment can also be used as a handy tool, for example, when maintaining genetic stocks with insufficient winter hardiness.

4.3 Plant Breeding and Plant Pathology

The breeder comes into contact with plant diseases in a variety of situations. Regardless of how well devised the genetic and methodical aspects may be, breeding programs will still be unsuccessful if there are great losses in the trial plots due to plant and animal pathogens. The breeder must therefore be able to take protective measures in time to prevent expected diseases or to effectively control their outbreak. Newly appearing diseases of an unknown type must be diagnosed by a phytopathologist, so that effective steps can be taken for their control. Furthermore, special attention must be paid to plant protection in the propagation of seed and plant material of varieties, as will be described in more detail later in Chapter 5.

4.3.1 Pathogen Diagnoses

When exporting seed and plant material, the often very strict regulations of the importing countries dealing with the proof that the import material is free of various pathogens must be observed. The same also applies for the exchange of small amounts of gene material for breeding purposes (genetic stocks). Vegetative material – such as wooden cuttings, tubers, bulbs – is often subjected to a lengthy quarantine, during which time it is thoroughly tested, especially for contamination with viruses and viroids, before it is released for use and further distribution. With regard to the prevention of spreading, especially of virus diseases, these regulations are justified. To test whether the material is free of virus diseases, special methods have been developed; viruses nowadays can thus be serologically detected in the pressed juices of the plants to be examined.

Since this method is a standard procedure, above all in breeding vegetatively propagated plants, the principle shall briefly be explained here. If suspensions of the viruses to be tested are injected into warm-blooded living beings, the

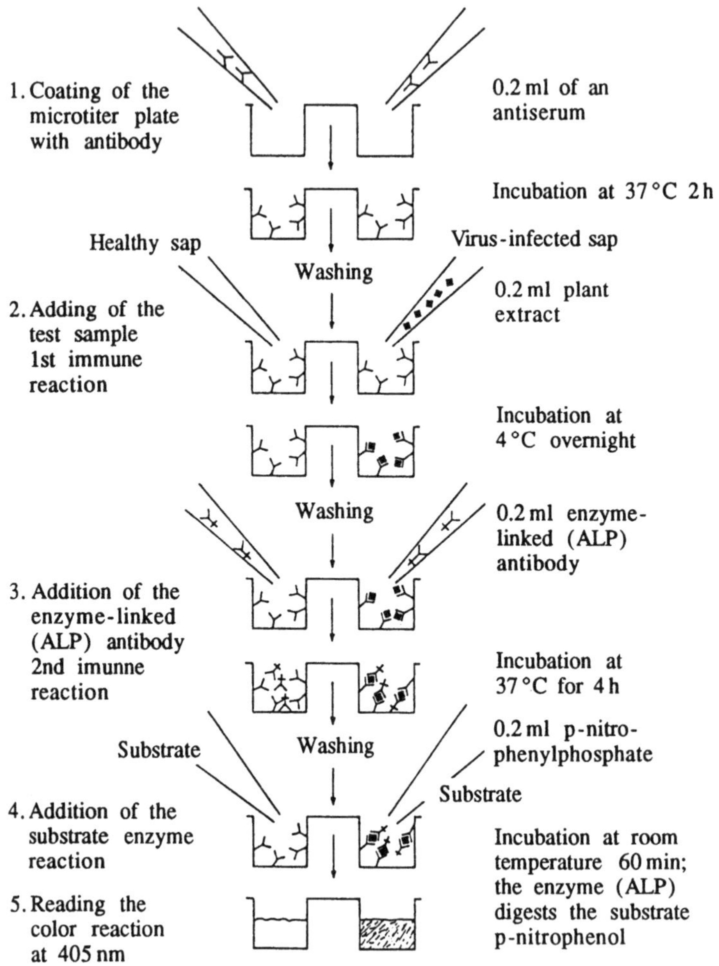

ALP = alkaline phosphatase

Fig. 61. Enzyme-linked immunosorbent assay (ELISA)

protein coats of the viruses work as antigens and induce formation of antisera, which can later be separated from the blood. The antisera contain antibodies that the warm-blooded being has developed against the virus protein, which specifically react later in vitro with the virus protein. This reaction is recognized as precipitation (precipitation test) due to agglutination of viruses in the juice (antigens) with the antibodies of the serum. In the meantime, this method has been further refined by linking the serological reaction to an enzymatic color reaction, by means of the *e*nzyme *l*inked *i*mmuno*s*orbent *a*ssay (ELISA, Fig. 61).

For the ELISA test, very small virus concentrations are sufficient. At present it is being considered whether the ELISA test can also be applied as a diagnostic tool in the identification of obscure symptoms of fungal or bacterial diseases. For the evidence of viroids that have no protein coat, hybridization with cDNA is increasingly used as proof.

Of particular interest for agriculture is the potato spindle tuber viroid (PSTV), a quarantine disease. After Van Wezenbeek et al. (1982) had developed cDNA probes and a routine hybridization technique, this method was extended to detect plant viruses. A further example is bacterial soft rot of potato tubers caused by *Erwinia carotovora*. The principle functioning of the test system is shown in Fig. 62. Currently, potato samples are checked by immunofluorescence, an immunological assay, or by ELISA. These tests cannot sufficiently differentiate between pathogenic and apathogenic *Erwinia* strains. Therefore, a DNA probe was developed that consists of the plasmid-coded region of one or more pectinolytic enzymes, which are responsible for the degradation of the cell wall and the subsequent rotting process.

Under applied aspects, it would be very beneficial to avoid the radioactive labelling of the cDNA probe. Success has been obtained for PSTV as well as for *Erwinia*. Although the sensitivity of biotin and ^{32}P-labelled probes is comparable (in both cases nearly 1 pg PSTV-RNA could be detected), specificity is still better in the radioactive system, which gives less false positives.

The two examples mentioned here illustrate that DNA probes can facilitate the rapid analysis of pathogens in plant tissues. As these tests are to some extent also quantitative, they prove beneficial for efficient resistance breeding

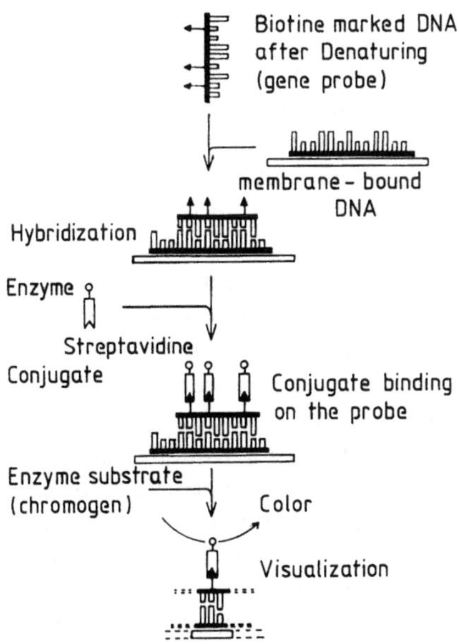

Fig. 62. Procedure for detection of pathogen DNA or of other DNA sequences using the cDNA hybridization technique

programs.If a virus disease of the plants is detected in corresponding investigations, it is possible to purify the plants by meristem tissue culture. To do this, pieces as small as possible are taken from the apical meristems (length 0.2 mm) of the infected plants (perhaps after a preceding heat treatment at 30–40 °C, a temperature that inhibits virus reproduction) and grown on culture media in vitro. A certain percentage of the regenerate plants is mostly free of virus and can be used as stock. In practice, meristem culture is of particular importance for vegetatively propagated crop species such as bulbous plants, potatoes, and fruit. Frequently, a propagation phase is directly initiated in vitro; for this combination of meristem culture as phytosanitary step and cloning, the term "rapid propagation" has come into use (see Sect. 5.3.1).

4.3.2 Breeding for Disease Resistance

The closest relationship between plant breeding and plant pathology exists in the field of breeding varieties that are resistant to pests and diseases. These breeding tasks generally require so much special knowledge and skills nowadays that they cannot be carried, if they are to have a chance of success, without the cooperation of a phytopathologist.

In the majority of our cultivated plants today, breeding of resistant varieties has become an important breeding goal. This also applies for diseases that can be controlled by fungicides and insecticides. Their application involves expenditures and occasionally, when improperly handled, the residues of the chemicals used can have a negative effect on the quality of the yield products; thus, the cultivation of a resistant variety is to be preferred.

The widespread opinion set forth in popular scientific publications that breeding for high yield is connected with an increasing susceptibility to diseases and that primitive and wild types are more resistant does not at all correspond to the facts. In many cultivated plants, varieties are known that combine high yield and resistance to diseases. Among the primitive varieties and the related wild species of our cultivated plants are also found susceptible genotypes. Population varieties are characterized by a greater genetic variability regarding resistance susceptibility, and are therefore an important source of resistance genes. Their detection within larger collections of primitive types is an important prerequisite for long-range resistance breeding, which is primarily combination breeding with special application of the effective backcross method.

Resistance is viewed, above all, as a factor of yield reliability, mainly because it is a preventive measure; it is never too late, as is often the case with spray applications. Aside from the aspect of yield reliability, a certain view is increasingly gaining acceptance, whereby expenses for spray applications can be saved by using resistant varieties. Table 24 shows the influence of diseases on the yield of wheat varieties with and without spray applications. The treatments for plant protection cause the most distinct yield increase in varieties that have only a low resistance level. In the variety "Kronjuwel", on the other hand, which has a high resistance level to the most important wheat diseases, foot, leaf, and ear treatment

Table 24. Yield comparisons of wheat varieties with different resistances without and with plant protection (data according to the local variety trial of winter wheat at the Agricultural Board Rhineland)

Variety	Behavior towards					Yield in t/ha		
	Mildew	Leaf rust	Yellow rust	*Septoria*	Mean	Without plant	With protection[a]	Difference
Kanzler	7	7	5	4	5.8	6.6	8.9	2.3
Caribo	8	6	5	4	5.8	7.0	9.2	2.2
Oberst	7	6	6	4	5.8	6.6	8.7	2.1
Monopol	7	5	5	5	5.5	6.9	8.5	1.6
Disponent	6	4	4	6	5.0	7.4	9.0	1.6
Carimulti	3	5	4	4	4.0	8.0	9.7	1.7
Granada	4	3	4	4	3.8	7.8	9.1	1.3
Kronjuwel	3	3	4	4	3.5	8.4	9.3	0.9

[a] Treatment of foot, leaf, and ear ; 9 = susceptible, 1 = resistant.

only result in a small yield increase. This variety almost reaches its maximum yield even without any spray application, a yield corresponding to the repeatedly sprayed variety "Kanzler". In addition to the total economic profit for the grower, the great importance of introducing resistance into new varieties from an ecological standpoint must also be pointed out.

In both stages of the breeding process – selection of resistant parents and selection of the sought-after combination types – methods of mass and early selection are used. After artificial infection in the greenhouse or in laboratories, selection can be carried out on a large scale. By controlling temperature, light, and atmospheric humidity it is possible to provide optimal conditions for infection and development of the pathogen and to systematically infect seedlings (e.g., late blight in potatoes, rust fungi in cereals, among others). Measurement under controlled laboratory conditions is an essential factor precisely under the aspect emphasized below that it is increasingly necessary to breed for quantitatively inherited horizontal resistance, for which quantitative test methods must be available. Only by eliminating as many noncontrollable influences as possible can reliable and reproducible data be obtained, which make possible the further setup of breeding through combination of several weaker resistances. Under the field conditions of the nursery, selection with the help of artificial infection becomes more difficult in many cases, however, this method is still to be preferred for achieving a particular type of resistance. In field trials it has proved effective to insert an especially susceptible variety at certain intervals between the plants to be tested and to plant it along the edges of the trial plots in order to spread the infection. As a rule, the susceptible variety is attacked by natural means. Otherwise, it is infected with spores of the pathogen that have been propagated in the greenhouse or laboratory. Spreading of the disease from the infected variety to the test plants is strongly dependent on the corresponding environmental conditions. Therefore, time of day and weather conditions are of decisive importance in artificial infection of the trial field.

The infection nursery should be set up in an area where the best natural conditions for the development of the pathogen prevail, so that a strong disease attack can be expected each year with great probability. Furthermore, the infection nursery must be isolated in such a way that by the provoked epidemic no other valuable stocks of the breeder are damaged that are not part of the resistance program.

The progeny of the plants selected as resistant are to be tested once more to guarantee the results of the artificial infection process. If there are segregations, single plants must be selected again for progeny tests under infectious conditions. After breeding to constancy, additional trials of the resistant lines take place for the evaluation of yield, quality of yield products, ecological variation, and other factors. These trials can be reduced to a minimum if the resistant lines have been developed from multiple backcrosses with an established standard variety.

4.3.3 Epidemiology

Along with increasing distribution of a resistant variety and simultaneous decrease in susceptible varieties, an infestation of the new resistant variety is often observed after a few years of successful cultivation, leading to a complete breakdown of the resistance upon further cultivation. The reason for this is the appearance of a new virulence of the pathogen in question, which is also characterized by a specific, genetically conditioned adaptation, so that it is able to infect the previously resistant variety. The breeder now begins – and usually not without success – to look for a new gene, the alleles of which, either in the recessive or dominant state, prevent infestation by the new virulent race of the pathogen. In general, the variety equipped with this new resistance gene will soon meet the same fate as the first resistant variety: the effectiveness also of this resistance gene will be overcome by a new virulent race and the variety will be infested. The breeder's race for new resistance genes begins anew, followed soon thereafter by the development of a new race, also equal to the new resistance gene as a result of further virulence. There appears to be a corresponding virulence gene in the parasite for every resistance gene of the host plant.

There are many examples of these relationships between host plant and pathogen in resistance breeding. From the numerous resistance genes used in breeding *Phytophthora*-resistant potatoes, resistance genes R_1, R_2, and R_3 and recessive gene r for lacking resistance are shown in Table 25 with their behavior in relation to races (0), (1), (2), (3), (1,2), (1,3), (2,3), and (1,2,3).

Race (0) can only attack varieties without the resistance gene, i.e., r genotypes. Race (1), on the other hand, also infects varieties with resistance gene R_1, but not with R_2 or R_3. Races (2) and (3) behave correspondingly, attacking

Table 25. International nomenclature of the relations between resistance genes and races of *Phytophtora infestans* (After v.d. Plank 1982)

Race Genotype host	(0)	(1)	(2)	(3)	(1,2)	(1,3)	(2,3)	(1,2,3)
r	S	S	S	S	S	S	S	S
R_1	R	S	R	R	S	S	R	S
R_2	R	R	S	R	S	R	S	S
R_3	R	R	R	S	R	S	S	S
R_1R_2	R	R	R	R	S	R	R	S
R_1R_3	R	R	R	R	R	S	R	S
R_2R_3	R	R	R	R	R	R	S	S
$R_1R_2R_3$	R	R	R	R	R	R	R	S

R = resistant; S = susceptible.

R_2 or R_3 varieties. There are still further races with combined virulence, such as (1,2), (1,3), (2,3), and (1,2,3). Race (1,2) infects varieties containing R_1 as well as R_2. The corresponding behavior of races (1,3) and (2,3) can easily be inferred from the given hypotheses. Now the exceptionally virulent race (1,2,3) is able to attack all genotypes of the host plant listed in the table.

The resistance of the host plant, differentiated by the varied interaction between genotypes of the host plant and races of the pathogen, is termed "vertical" or specific resistance. According to v.d. Plank (1982), vertical resistance of varieties and races with a differing virulence are two aspects of one and the same phenomenon. The comments that follow are essentially based on the facts and their interpretation regarding vertical and horizontal resistance in the book by v. d. Plank. Vertical resistance is often simply inherited if it is monogenic or oligogenic, i.e., a discontinuous variation is observed in a cross analysis, making it possible to form classes with regard to resistance or susceptibility. Phenotypically, vertical resistance mostly manifests itself in hypersensitivity of the host plant. If the germ tube of the spores of the pathogen penetrate the cells of the host plant, the cells die off; with this defensive reaction, the basis of life is also withdrawn at the same time from the parasite. The appearance of new races of the pathogen with an increased virulence can be explained by selection processes within the population of the pathogen. If susceptible varieties are cultivated, the population of the parasite chiefly consists of (0) races. If a resistant variety R_1 is now cultivated, it cannot be attacked by the (0) race, which is reduced to a minimum proportion of the population within a few years, while simultaneously a more virulent race (1) multiplies to such an extent that the population is mainly composed of this race, and thus the resistant variety falls victim to it, too. New virulent races are formed through mutation and recombination, insofar as they have not already been present in the population.

Besides these races characterized by differing virulence, there are also those that cannot be distinguished by different varieties of the host plant. Here, there is no interaction between races and varieties as to sequence, as can be seen from the case in Table 26. The tomato varieties "Bonner Beste" and "Maryglobe" were

Table 26. Behavior of five races of *Fusarium oxysporum* towards two tomato varieties (After v.d. Plank 1982)

Races		Strength of reaction "Bonner Beste"	"Maryglobe"
I	Elevated loose	6.2	4.5
II	Elevated sclerotic	5.2	3.8
III	Medium elevated	5.0	2.7
IV	Medium plain	3.8	2.5
V	Plain moist	2.8	1.8

1 = Not affected; 9 = very heavily affected.

infected with five races of *Fusarium oxysporum*, which are characterized by differences in their mycelium growth. The host reaction to the infection is rated on a scale, with 1 signifying no disease symptoms at all and 9, very strong symptoms. All five races attack both varieties, but with a different intensity: "Maryglobe" is more resistant than "Bonner Beste." Thus, the differences in resistance are of a quantitative nature. As to the varying aggressiveness of the five races, they display on both varieties the same sequence after infection. In this case, no differentiation of dissimilar physiological races is possible by the infection of two genotypes of a host plant. However, the five races of *Fusarium oxysporum* can be distinguished by their differing reaction on *one* variety only, either on "Bonner Beste" or "Maryglobe." The different degree of infestation of the races is also of a quantitative nature. The races show no specific adaptation to certain varieties. The varying behavior of different races of a parasite on one variety is termed aggressiveness and the resistance, different as to varieties but unspecific as to races, is termed "horizontal" or unspecific resistance. As a rule, horizontal resistance is polygenically determined, i.e., after cross with a susceptible variety there is a continuous variation in the degree of susceptibility/resistance in the F_2; formation of classes, as is often typical for monogenic or oligogenic segregations, is not practicable.

Horizontal or unspecific resistance phenotypically manifests itself, among other things, in the fact that the number of spores needed to induce an infection is considerably larger than in a susceptible variety, whereas the number of pustules formed is smaller and the period of latency until formation of new spores is distinctly prolonged. While vertical resistance delays the outbreak of an epidemic, spreading of infection can be slowed down so much in varieties with horizontal resistance – especially in varieties with a long vegetation period – that no greater infestation whatsoever occurs and thus no decrease in yield.

The differences between both types of resistance shall once more be summarized as follows: vertical resistance is differentiated as to the races of the pathogen; horizontal, on the other hand, is uniform and independent of the races. The differences between varieties are qualitative in vertical resistance, in horizontal they are quantitative. In vertical resistance, the inheritance is monogenic or oligogenic, whereas in horizontal resistance it is often polygenic.

Strongly virulent races, such as race (1) of *Phytophthora infestans*, which also infests potato varieties with the resistance gene R_1, can also attack varieties without any resistance gene, but can be inferior to the race (0) attacking there simultaneously. The high virulence of race (1) has an adverse effect when competing with race (0) for attack of the variety without R_1, so that the race with the stronger virulence is eliminated again from the pathogen population by further cultivation of the variety without the R gene. After cessation of the selection pressure resulting from the resistance gene, the race population has the same composition as it had before the introduction of the variety with the resistance gene.

An example of low aggressivity in higher virulence is the result of two races of black rust in wheat. Race 56 and race 15 B_1 both attack the wheat varieties "Little Club, Marquis", and "Red Bob". Race 15 B_1 additionally attacks further varieties such as H 44, Mindum, and others that are resistant to race 56, i.e., race 15 B_1 is more virulent than race 56. If pustule formation of both races on the varieties infested by both races is compared (Table 27), it becomes clear that the more virulent race 15 B_1 is less aggressive – apparent by the number of pustules – than race 56. The less virulent race 56 is thus superior to the common host, e.g., these three varieties.

The example cited also makes it clear, however, that the characterization of virulent and aggressive races is dependent on the type and size of the test material used. To the same extent, "vertical resistance" and "horizontal resistance" can both be seen as extreme concepts, between which the resistance characters of the varieties lie.

The genotypes of the host plants determine the changes in the genetic composition of the race population. The vertical resistance of a variety can only

Table 27. Different virulence and aggressiveness of two races of *Puccinia graminis* (v.d. Plank 1982)

Races	Varieties				
	Little Club	Marquis	Red Bob	Hope 44	Mindum
	a) Differences in virulence				
56	S[a]	S	S	R	R
15 B1	S	S	S	S	S
	b) Differences in aggressiveness (number of pustules)				
56	6,292	4,374	3,594	–	–
15 B1	5,448	3,065	1,646		

[a]R = resistant; S = susceptible.

survive as long as the specific virulent races are eliminated by the stabilizing selection of a susceptible variety. A pertinent example for this is the cultivation of spring wheat in the northern regions of the USA, the black rust resistance of which is based on gene S_{r6}. After the eradication of barberry, the obligate parasite does not have any chance of surviving the winter in the north, so that only the spores coming from the south every year can induce an infestation. As long as spring wheat without resistance gene S_{r6} was cultivated in the south, the races attacking S_{r6} varieties were eliminated again because of their smaller adaptability within the race population, and the vertical resistance of northern spring wheat was maintained. However, as soon as spring wheat varieties with gene S_{r6} conquered cultivation in the south, the compatible races were able to multiply so strongly that the resistance of the northern spring wheat varieties collapsed. In order to maintain vertical resistance, the behavior of the host plant/parasite is not only of prime importance, but also the reaction between *two* genetically different host plants and the parasite.

In facultative parasites with a saprophytic phase, the filtering effect of a host plant without a resistance gene, which in obligate parasites leads to the elimination of races with an unnecessary virulence, is taken over by the saprophytic substrate, e.g., root residues in the soil.

4.3.4 Strategies for Durable Resistance

Which resistance should the breeder make use of in order to attain his breeding goals? In past decades, vertical resistance has come into the forefront of efforts in breeding. The methodology can be relatively simply handled: backcrossing combined with mass infection and early selection in the greenhouse with controllable environmental conditions. Great successes were achieved, but the setbacks were also just as great.

As a rule, vertical resistance promises prolonged success if strong resistance genes are used and if a nonspecific resistant host variety is inserted, by which the specific virulent race of the pathogen is again weeded out due to lack of aggressivity. This stabilizing selection of the parasite population does not take place in races having a correspondingly smaller virulence to overcome a weak gene and thus also a greater adaptability to host plants without resistance genes, or in nonobligate parasites in the substrate of the saprophytic interval.

If the resistance is based on a strong gene with dominant inheritance, it can successfully be used in the development of resistant hybrid varieties. An example for this is the tomato F_1 hybrid "Immuna", which was developed in Weibullsholm, Sweden, and is resistant to the brown spotted disease, *Cladosporium fulvum*.

Strong vertical resistance genes can be used with great success in the creation of so-called "multiline" varieties (Borlaug 1958). Genotypes with different resistance genes are backcrossed several times with a standard variety. In this way, lines are formed which resemble the standard variety in all characters important for cultivation and utilization, but which are distinguished by different resistance genes. One variety is composed of three to five such lines that are, e.g., resistant to different races of yellow rust in wheat. The advantage of such a variety is that reproduction of some races of the pathogen population is inhibited by specific resistance genes of the lines, and thereby spreading of an epidemic is checked. Particularly virulent races may be able to overcome the resistance of certain lines, but are again reduced in their numerical proportion of the parasite population by lines in which such a strong virulence is not necessary for infestation due to lack of adaptation. For the grower, a multiline variety represents *one* host plant variety, but such a variety contains several genotypically different host plants for the interaction between parasite and host plant, so that none of the races can have mass propagation. Consequently, multiline varieties provide better conditions for the maintenance of a vertical resistance. Nevertheless, trials with multilines of barley with powdery mildew infestation have shown that complex virulences accumulate surprisingly quickly. Multilines can therefore not be regarded as equivalent substitutes for varieties with horizontal resistance.

Vertical resistance cannot be used in perennial plants, because with the possibilities presented here of continuous multiplication of the pathogen, there is no stabilizing selection by other host plants or through the saprophytic phase. Furthermore, vertical resistance is not possible if no resistance gene to an especially virulent race has yet been found in the host plant. Here, breeding for horizontal resistance could be an alternative.

Circumstances regarding the effectiveness of a vertical or horizontal resistance in varieties with long vegetation period could be quite similar to those of perennial species. In short-lived, early varieties of potato, a vertical resistance to *Phytophthora infestans* initially delays infestation. Before a more virulent race has reproduced to an extent that it could endanger the early variety, the vegetation period for this variety is already finished. The vertical resistance prevents a decrease in yield of an early variety. In a late variety with vertical resistance, the more virulent race can strongly propagate, so that an epidemic infestation can cause considerable losses in yield. Late varieties with horizontal resistance are attacked earlier than varieties with vertical resistance, but reproduction is slowed down by the horizontal resistance to the point that an epidemic cannot spread. Consequently, in varieties with a long vegetation period, horizontal resistance is more effective.

If selection for vertical resistance is deliberately carried out in an extensive breeding material for many years, the chances of also successfully selecting for horizontal resistance in the same material later are inevitably diminished. In breeding for vertical resistance, a large portion of the cross material has already been eliminated as seedlings by artificial infection; in the remaining material, resistance genes prevent infestation so that simultaneous selection for horizontal resistance is not possible. Polygenes for horizontal resistance can be lost through one-sided selection for vertical resistance; at the very least, there is no accumulation of such genes in the individual genotypes. However, the art of breeding for horizontal resistance is precisely to increase resistance by the purposeful combination of a large number of genes with a small effect (polygenes).

In cross-pollinated populations or varieties such as maize, an accumulation of genes or alleles for horizontal resistance necessarily takes place because of the selection pressure of the parasite. The survival chance of a genotype increases with the number of favorable polygenes. On the other hand, the horizontal resistance of a population is lost as soon as selection pressure via the parasite diminishes or disappears completely. Then the alleles of the relevant gene loci, which do not have any effect on resistance, probably have a selective advantage. The lost resistance can be regained with newly begun distribution of the pathogen.

It is more difficult to breed for horizontal resistance in self-fertilizing crops such as barley, wheat, oats, and others. Here, it is recommended that bulks be formed from a large number of single crosses, so-called composite crosses, to generate original material with as large a genetic variation as possible, on which the pathogen can have a selective effect. By use of male-sterile genotypes and the cross-pollination linked thereto, the possibility exists of forming new recombinations also in later generations of the bulks, and to thereby broaden the selection basis for horizontal resistance.

An improved trial method for horizontal resistance to *Phytophthora infestans* in potato has been developed by Umaerus (1969) and successfully applied in selection. Quantitative gradation in resistance could initially be established by spraying potato leaves in petri dishes with a standardized zoospore suspension of fixed concentration. The infection process was then abruptly interrupted after a limited duration of ca. 9 h by drying the leaves. As a quantitative measure of resistance, the quotient of the number of infected spots (lesions) per area unit in relation to the number of applied spores was used and was designated IEv-index (Infection Efficiency of variety). This quantitative method had proved to be a success in the infection of 6-week-old seedlings in the four- to five-leaf stage (infections to vertical resistance are already carried out in the seed-leaf stage). It is a prerequisite that temperature and atmospheric humidity be kept constant and that there is a quick transition to the post-infection conditions by drying, and also that the pathogen races used are virulent enough to overcome the effectiveness of all R genes contained in the trial material, i.e., it precludes the presence of vertical resistance of the seedlings.

After lines with a quantitatively graded horizontal resistance have been found, the next step in breeding is to cross such mutants in expectation of obtaining transgressions in the polygenic horizontal resistance. The considerably greater difficulties in breeding varieties with horizontal resistance are compensated for by the prospect of lasting success. It is very probable that transitional types also occur between vertical and horizontal resistance. When setting up a breeding program and the methods to be used, the breeder must first proceed with a clear conception of the desired or achievable type of resistance, the best help for which are – as extreme terms – the described alternatives "vertical or specific" resistance, on the one hand, and "horizontal or unspecific" resistance, on the other.

Finally, the possibility of breeding "tolerant" varieties must also be pointed out. These are varieties that only have small yield losses or decreases in quality despite strong infestation. However, the danger that the pathogen can uncontrollably multiply and spread in the cultivation of such varieties should not be underestimated.

It has been the purpose of this chapter to make clear to the future breeder how important it is to acquire basic knowledge in the fields of plant production in general and plant physiology and pathology in particular, as a foundation for effective cooperation. In these fields, specialization is so far advanced that a breeder can no longer be proficient in all of them.

4.4 Breeding of Industrial Crop Species

Since ancient times plants such as flax or cotton have been cultivated, the products of which have been used for nonfood purposes. The most significant plant group used in industry are, however, surely trees for wood production. But their importance is so large and the breeding strategies are so different that tree breeding – neither for forest trees nor for fruit trees – will be covered in this book. More recently, regular industrial crops such as oil palm or rubber tree have become very important. In those countries in which the cultivation of industrial crop species is of special economic importance, plant breeding activities of these species have always played an economic role. In contrast, in many industrialized countries of northern temperate climates, this branch of plant breeding came to a halt some decades ago, the products of the chemical industry pushing aside plant production. In these countries, meanwhile, a reassessment has taken place. The reason for this is the perception that mineral oil, as a versatile raw material, is not inexhaustible and, moreover, that many plants can supply prefabricated first-rate industrial-base materials, e.g., fatty acids in seed oil or medicines as secondary metabolites. In this chapter, therefore, the term "breeding of industrial crop species" refers exclusively to this new development in agriculture.

Under this premise, the following products are basically of interest: oil and fats, starch, sugar substances, fiber, proteins, dyes, waxes, and pharmaceuticals. Important crop species of these groups are listed in Table 28.

From this table it can be recognized that not only well-known crop species of today belong in this group, but also semidomesticated and wild species. Accordingly, several approaches to industrial crop plant breeding are feasible:

1. A reutilization of present varieties, i.e., a modification by breeding to gain improved suitability. Example: erucic-acid-free rape for the food industry versus erucic-acid-containing rape for oleochemistry.
2. A revival of old obsolete crop species, where an increase of yield to the actual level of comparable crop species is necessary. Examples: *Linum usitatissimum* L. (flax as well as linseed), false flax [*Camelina sativa* (L.) Crantz.]
3. Domestication of new species from the wild flora or of already semidomesticated species. Examples for wild species: caper spurge (*Euphorbia lathyris* L.), meadowfoam (*Limnanthes* spp.), or *Cuphea* spp. for seed oil of special quality. Examples for semidomesticated species: a number of umbellifers (*Apiaceae*) hitherto grown as spices and in many cases as medicinal plant species.

Table 28. Important industrial materials and the corresponding plant species (selected)

Seed oil	*Crambe*, caper spurge, *Cuphea* spp., umbelliferous species (Apiaceae), meadowfoam, *Lesquerella* spp., purple fleabane, jojoba, common marigold, linseed, false flax
Starch	Potato, peas, lupin, *Phaseolus* beans
Sugar substances	Beets, sweet sorghum, chicory, Jerusalem artichoke
Fibres	Hemp, flax, nettles
Pharmaceuticals	Foxglove species, *Belladona*, poppy, yellow bark, coca, Dalmatian insect flower (Just a few of a vast number of additional species)

In any case, comprehensive screenings and a thorough evaluation of available germ plasm as to agronomical aspects and intended use is an essential prerequisite. In the last few years, major efforts in various countries were directed towards seed-oil-containing species from the wild flora. The model for this comes from the USA. The US program for new industrial plants from wild species began in the second half of the 1950s. It was mainly oriented towards evaluation of new oil crop species. The basic purpose for initiating this research work was: the development of labor-intensive, special cultures for areas of high population density, to realize a higher income than can be attained by growing common bulk goods; the inclusion of dry areas or, for other reasons, marginal sites to produce market goods; and the assurance of the supply by supplementing the petrochemical raw material or other tropical oils.

It appeared that many seed oils exhibited an unusual and even unknown composition of fatty acids that are of economic importance. A selection of the more important breeding problems involved can be illustrated by the following case studies of certain species.

Crambe (Crambe abyssinica Hochst. ex. R.E. Fries)

This species was a new introduction for the United States. The first screenings confirmed a lack of genetic variability for important agronomic traits among the initial accessions. Consequently, the need to acquire additional germ plasm exhibiting natural variation or to create new variability through mutagenesis seemed to be inevitable.

The initial conclusion that Crambe showed inadequate variation was the result of a few accessions and a lack of information on cultivation techniques for testing and on plot size and shape. After crossing appropriate strains and revising plot techniques, it was possible to select improved strains.

Table 29. Plant characteristics of *Euphorbia lathyris* and their inheritance

Character	Wild type	Crop plant	Mode of inheritance
Life cycle	Biennial	Annual	Monogenic
Seed dormancy	Strongly marked	Missing or weakly marked	Oligogenic (?)
Seed dispersal mechanism	Present	Absent	Oligogenic
Flowering and ripening period	Extended	Concentrated	Polygenic
Toxic constituents	Present	Absent or reduced content	Polygenic (?)

Caper spurge (Euphorbia lathyris L.)

Another example of a species involving special breeding problems is caper spurge. Although cultivated in the past for medicinal purposes and also grown nowadays in gardens, this species maintained typical wild-plant characteristics. Table 29 gives an overview of this species, including the mode of inheritance. While both annual genotypes and indehiscent fruits can be obtained easily with this partially allogamous species, quantitatively inherited traits are more difficult to work with, particularly the long-lasting flowering and ripening period resulting from an indeterminate inflorescence. In this regard, presumably the application of special growth retardants will succeed in a more concentrated flowering and ripening time. A special problem is the occurrence of toxic constituents, i.e., diterpene esters, which are undesirable even in oleochemistry. To keep this under control, a speedy screening method must be worked out, so that large numbers of individuals can be tested in the search for genotypes of low or zero diterpene ester content. It also transpired that accessions supplied by botanical gardens displayed only modest variability with regard to yield-determining traits in contrast to that of genetically divergent populations. These are certainly small isolated populations of limited intraspecific variation, which apparently were inbred for generations. Broadening the genetic base of this species, therefore, is an urgent prerequisite before significant progress can be made.

Cuphea species

Because of its high content of medium-chained fatty acids in the seed oil this genus, consisting of perennial and annual species, calls for special attention. A comprehensive investigation on interspecific variability of 18 species in six taxonomic sections showed promising results. The fatty acid spectrum of various species, however, exhibited a low level of intraspecific variation. All species showed typical wild-plant characters, among these, rupturing of ripening fruits

Table 30. Wild-type characteristics of *Cuphea* and variants usable in breeding (Röbbelen 1982)

Character	Genetic variation
Hard-shelled seeds and seed dormancy	Quantitative variation
Dorsiventral growth habit	Monoculm mutant
Viscid glandular hairs	Mutant without glandular hairs
Rupturing fruits	Mutant with radial symmetric flowers
Long-lasting flowering and irregular ripening	Fasciata mutant and quantitative variation

attributable to zygomorphy of flowers as a generic character. Therefore, mutation experiments were carried out to obtain mutants of some breeding value (Table 30). Due to pleiotropic gene action in some mutants, disadvantageous effects appeared. Moreover, attempts had to be made to increase the expression of mutant genes by transferring them to another genetic background. Autogamous *Cuphea* species rate higher than allogamous with respect to future prospects.

These few examples of seed-oil-containing wild species clearly demonstrate that use of such species demands different breeding approaches if the prospects are to be promising.

Again, other breeding problems can be demonstrated by citing certain fiber crops. Growing fiber crops is one of the early achievements of civilization, with flax as the classical example. Nevertheless, cultivation of flax in industrialized countries came to a standstill after World War II due to industrial fibers being produced from mineral oil. As late as the 1960s comprehensive screenings were put into operation. Thus, in the United States alone, 387 mainly herbaceous species from 44 monocot and dicot families were evaluated with respect to their utilization, above all, for papermaking.

The fiber content of dioecious hemp (*Cannabis sativa* L.) was increased within 8 years in selected female plants, from 13.1 to 22.7% on the average, and from 10.2 to 14.7% in selected male plants (Bredemann 1942). However, a significant breakthrough towards a fiber-rich hemp containing an average of 20% fibers followed from breeding monoecious genotypes (Bredemann et al. 1961). This advance has still not been made with the similarly dioecious nettle (*Urtica dioica* L.), the fibers of which are used for technical purposes. Plantations of fiber nettle up to now have been and still are composed of easily obtainable clones consisting of either female or male plants.

Linum breeding, especially flax, has never been given up entirely and a sufficient variability in varieties and landraces is still available. There are no special problems; basically, breeding is directed towards yield increase and stability.

A quite different situation exists in medicinal plants and spices. On the whole, there are several thousand plant-derived pharmaceuticals that have medicinal applications. Well-known drugs are: *Digitalis glycosides* for oral contraception, anabolics, and corticosteroids; *Belladonna* alkaloids (codeine, morphine) as parasympatholytic agents; opium alkaloids (codeine, morphine) from *Papaver somniferum* L. as analgesics; quinidine and quinine from *Cinchona* sp. as antiarrhythmics, and antimalarial agents; and cocaine from *Erythroxylon coca* as local anesthetics. Even pesticides derived from "insecticidal" plants are in common use, pyrethrin from Pyrethrum (*Chrysanthemum* spp.) being the best-known example. Because the vast number of chemical compounds are nearly exclusively secondary metabolites, and because of the restricted growing area of a single plant species, breeding activities regarding these species are very limited. In most cases, therefore, mass selection within populations has led to the varietal structures. Modern breeding methods are still the noted exception.

4.5 Plant Breeding for Developing Countries

After the discovery and description of natural diversity in the "gene centers" by N.I. Vavilov in 1928, the possibilities of utilization and the necessity of safeguarding these resources were first realized in those countries that practiced plant breeding. In countries where populations of great diversity were still present, i.e., mainly in the developing countries, more intensive plant breeding activities were started rather late.

The plant populations grown in these countries were adapted to the environmental conditions of growing areas by long-standing natural selection. If natural populations are still grown in such a country, they should be utilized as a base population for the selection of lines. At the same time, crossbreeding should be initiated in order to combine the adaptability of selected lines with the yield potential, quality characters, or resistance of introduced genotypes.

Looking at the great number of cultivated crop plants available in the developing countries, it is obvious that many breeding problems existed that had to be solved. Increasing population pressure coupled with socioeconomic problems required improved plant productivity per acreage. When in the mid-1960s so-called high yielding varieties (HYV) of wheat became widespread throughout cereal-growing areas of the world, a general interest to introduce them arose. These varieties were the first great success of the breeding work of CIMMYT (see Table 31 for abbreviations). The capacity for performance and adaption of HYVs is the result of combining the traits of genetically divergent cross parents. The HYVs represent semidwarfs exhibiting a high lodging resistance, long ears, and multiflowered spikelets and, moreover, a high response to increased fertilization and irrigation. Their adaptability is based on low photoperiodic reaction (photo-insensitivity), in which the yield formation is not negatively influenced when grown under varying day lengths.

Countries with the appropriate agronomic conditions showed great success in growing these varieties. As early as 1970 wheat yields in India, for instance,

Table 31. International Agricultural Research Centers running plant breeding activities

A) Institutes under the auspices of the Consultive Group of International Agricultural Research (CGIAR)	
CIAT	(Centro Internacional de Agricultura Tropical; Cali, Colombia): *Phaseolus* beans, cassava, forage grasses
CIMMYT	(Centro Internacional de Mejoramiento de Maiz y Trigo; Mexico DF, Mexico): wheat, maize, barley
CIP	(Centro Internacional de la Papa; Lima, Peru): potato, sweet potatoe
ICARDA	(International Center for Agricultural Research in the Dry Areas; Aleppo, Syria): wheat (*durum*), barley, *Faba* bean, lentil, chickpea, *Medicago*
ICRISAT	(International Crops Research Institute for the Semiarid Tropics; Patancheru, Andra Pradesh, India): *Sorghum*, millets, chickpea, pigeon pea, groundnut
IITA	(International Institute of Tropical Agriculture; Ibadan, Nigeria): groundnut, soya, sweet potato, cassava, cowpea, rice
IRRI	(International Rice Research Institute; Los Baños, Philippines): Rice
B) Other Institutions	
FAO/IAEA:	Joint FAO/IAEA Division of Nuclear Techniques in Food and Agriculture, Vienna
GSF:	Gesellschaft für Strahlen- und Umweltforschung, Munich-Neuherberg
IAEA:	International Atomic Energy Agency, Vienna
SAREC:	Swedish Agency for Research Cooperation with Developing Countries, Stockholm
SIDA:	Swedish International Development Agency, Stockholm

increased by 62.6% and in Pakistan by 82.6% from 1965 yields. There was the belief in a "Green Revolution" that could help overcome the worldwide hunger problem. But very soon serious drawbacks were reported from areas where basic requirements could not be assumed a priori. In those areas yields were lower than that of indigenous varieties and populations. At least as serious was the loss of genetic resources following the displacement of indigenous varieties and landraces. Nevertheless, despite many of the negative consequences for traditional farmers, it must be stated clearly that this new agricultural technology has no built-in source of social discrimination as such.

The setbacks prompted CIMMYT to change its policies in that more and more advanced populations of semidwarfs were introduced that originated from crosses tested for combining ability. In India, for example, as early as 1963 selection in such introduced populations under different environmental conditions led to four varieties which, due to their high adaptability, found a wide distribution. Furthermore, from crosses of introduced semidwarf genotypes with long-stemmed, lodging-susceptible Indian varieties, some 15 varieties were created. Yet the basic requirement for success in achieving high yields was still improved cultivation technique and increased application of fertilizers.

Another example that illustrates the importance of indigenous adapted populations for selection, followed by outcrossing selected lines to introduced high-yielding varieties, comes from a wheat breeding project of the Food and Agriculture Organization (FAO) in Iran from 1952 to 1968. Starting with 7000 progenies from single spikes that were selected from some 440 populations or local varieties, four uniform varieties were released, all outyielding the initial material by 18–73%, depending on variety and growing area.

Together with the selection of promising genotypes, an outcrossing project was started using lodging resistance, disease resistance, and baking quality and having foreign varieties as the second cross parent. Only those combinations proved to be successful in which one parent was of Iranian origin. In the lowlands as in the Caspian Sea area, Mexican cultivars outyielded Iranian ones derived from crosses: 3.9 tons/ha versus 3.4 tons/ha; in the highlands the Iranian varieties, yielding up to 4.6 tons/ha, were superior to the Mexican ones, if fertilizers and sufficient irrigation were used in advanced agricultural production. Without a doubt, in this case a primarily yield-oriented plant breeding strategy has had favorable effects on the nutritional well-being of the population.

Another substantial contribution to the so-called Green Revolution was made by IRRI. In 1966, 4 years after starting research, this institute released the high-yielding, early-maturing semidwarf variety, IR8. In the following years, new cultivars resulted from numerous crosses, mainly distinguished by an improved grain quality as well as by increased resistance against pests and diseases. IR36, the pedigree of which includes the wild species *Oryza nivara* for enhancement of resistance, proved to be of particular value. This cultivar occupied 11 million hectares in 1982, about 7% of the worldwide growing area, thus being the most widespread rice variety. Another cultivar, IR58, released in 1982, is characterized by an extremely short vegetation period. Ripening 10 days earlier than IR36, it allows more than one harvest per year. On a worldwide scale, 155 breeding lines of irrigated rice were established as named varieties (IRRI 1987). One of the most remarkable achievements in rice breeding, however, was the recent release of an apomictic cultivar in mainland China.

This new cultivar combines high yield, good tolerance in stress, and sequential use of heterosis. The tropical cultigen, cassava (*Manihot esculenta* Crantz), now widely grown for its starchy roots, is another outstanding example where national and international efforts improved agriculture in developing countries. Breeding experiments with this vegetatively propagated species, primarily a small-farmer crop, started in the twentieth century in Africa, Asia, and Latin America. National institutes made notable progress up to 1960, when the species entered into international agricultural research at IITA (see Table 31 for abbreviations). A large part of all cassava breeding is devoted to resistance against pests and diseases, since the clonal plants are especially endangered. As to the progress in root yield, both the data from experimental stations and from farmer's fields confirm the superiority of new hybrids over local varieties. In Latin America, yields of 2 tons/ha on soils of low to moderate fertility are reported for new cultivars, and in Africa even more. Various international agricultural research institutes work with a number of other important field crops. Beans, potatoes, fodder crops, grain legumes, sorghum, millet, groundnuts, and sweet potatoes are the best-known examples. A common feature of this research and developmental work is the close cooperation of the international institutes with national ones located in the crop-specific regions. Meanwhile, also modern breeding techniques besides conventional ones such as haploid production or in vitro cultures, etc., have been incorporated into the programs.

Another way plant breeding was promoted in the developing countries was by coordinated projects. These were executed by the special institutions or agencies cooperating in the work, which particularly addressed problems of great technical requirement.

Besides developing inbred pure lines or open-pollinated varieties, which includes also the use of induced mutants for broadening the genetic variation of cross parents, hybrid varieties play a more important role for plant breeding in developing countries. As a consequence of greater efficiency in the breeding business, but initiated more by international cooperation, the technical requirements for the expensive development of inbred lines and for the production of hybrid seed were provided. A classic example is the growing of hybrid maize in Kenya. This was promoted by an agreement between the governments of Kenya and The Netherlands in 1971. About 800 000 bags of hybrid maize seed of 10 kg each were retailed to small farmers in 1974/75. The success of the project depended very much on the availability of good, locally bred maize varieties, on the management of a seed inspection service and, last but not least, on an active partnership of seed trade. As early as 1976/77, the Kenya Seed Company produced 13 000 tons of hybrid maize seed. Also the seed certification of bred varieties was introduced. After 5 years, certified seeds comprised more than 40% of the total growing area. Only by producing and growing high-quality seed of productive varieties can plant breeding make an economic contribution.

Certainly the regulations for variety acceptance and seed certification are not of the same standard as has been set by the 1961 Paris Convention for European countries, but prospects for amendments are good. A report of Rohrmoser and Seidel (1977) contains abundant information for the promotion of seed production in the developing world.

Many projects clearly demonstrate the need of a more nutritionally balanced diet. If, therefore, high yield and improved nutritional quality could be combined, it would obviously be beneficial. Instead of primarily cereal-based food, however, a combined legume-cereal diet promotes more balanced nutrition. In the case of tuber crops such as cassava, a suitable intercropping system with a grain legume would help to achieve a simultaneous improvement in energy and protein yield. Furthermore, yield stability is a prerequisite for successful small farming.

Meanwhile, the breeding scene in the developing countries has changed. More international institutes, which as nonprofit organizations are engaged in plant breeding activities on a worldwide scale, and also national plant breeding institutes have been established and provided with modern scientific equipment. In some of the developing countries, a central authority aided by regional experimental stations is in operation. In general, there is a rather close cooperation with the corresponding international institute by means of joint ventures for mutual benefit. Moreover, due to these efforts, the education and training of scientists and technicians, who are able to apply modern plant breeding methods and are willing to do the indispensable fieldwork, have made considerable progress.

Modern plant breeding involves high investment costs, which pay off only by increasing seed returns. Small seed companies, therefore, are becoming less competitive compared to the large ones. In recent years, multinationals of the chemical and pharmaceutical industry have very often taken over seed companies. Combining mineral fertilizers, plant protection chemicals, and "homemade" plant varieties into one package ensures economic profit. Precisely in those countries where agriculture is intensifying plant production, a big market for mineral fertilizers, pesticides, and certified seed is on the horizon.

If such enterprises completely take over, the market runs a large risk that seed, especially of genetically uniform hybrid varieties of great adaptability, will find a wide distribution, the uniformity of which is subject to genetic vulnerability (see Sect. 6.1). Considering the great differences in climate, altitude, and soil qualities that are encountered in many developing countries, breeding of special cultivars certainly cannot be abandoned. In this field, great opportunity still exists for domestic enterprises and/or state institutions, also in cooperation with the international agricultural research centers. The intensified progress in plant breeding in Third World countries has also had an influence on the maintenance and utilization of plant resources. The global IBPGR network, in cooperation with the IARCs (see Table 31 for abbreviations), is of mutual benefit to the genetic resource activities and the plant breeders of these countries.

A close cooperation between plant genetic resource centers and breeding institutes is to be recommended, also in respect to the introduction of exotic crop species hitherto not cultivated in a particular country. There are many examples for importing crop species from one continent to another and of their successful distribution and utilization, both throughout history and recently. The transplantation of European cultigens to North America when this subcontinent was settled by European immigrants can be cited as just one example; another is the introduction of potato cultivation into Prussia in the eighteenth century, which was based on former introductions from South America to Europe. In more recent times, the distribution of industrial crop plant species such as caoutchouc, castor bean, and cacao from South America, or coffee from Africa into other countries and continents had a far-reaching impact on the importing areas.

It must be pointed out that new agricultural technologies, under which new crop species are to be classed, should be encouraged to develop in harmony with the socioeconomic needs of an area. Concerning the agroeconomic aspects, some important questions must be answered before a new crop species can be introduced:

1. How great is the adaptability of the prospective new crop species for the soil and climatic conditions of the host country with respect to economic considerations?
2. What special cultivation requirements are needed with regard to sowing date, sowing distance, fertilization, plant protection, etc.?
3. Which of the foreign varieties has the best adaptability?

4. Which preconditions exist or must be established for marketing the plant products? What are the domestic supplies? Is there the possibility of exporting these products to foreign markets?

Only if these problems have been solved satisfactorily and, furthermore, only if the right conclusions have been drawn can the results of plant breeding research be incorporated most favorably into the practical work of developing countries.

4.6 Cited and Recommended Literature for Chapter 4

Austin RB, Bingham J, Blackwell RD, Evans LT, Ford MA, Morgan CL, Taylor M (1980) Genetic improvement in winter wheat yields since 1900 and associated physiological changes. J Agr Sci 94:675–689

Baulcombe DC, Bouton RE, Flavell RB, Jellis GL (1984) Recombinant DNA probes for detection of viruses on plants. In: Brit Crop Protection Conf Pest and Diseases ed. Brit Crop Protection Council Publ, Croydon, pp 207–213

Borlaug NE (1958) The use of multilineal or composite varieties to control airborne epidemic diseases of selfpollinated crop plants. First Int Wheat Genet Symp, Univ Manitoba, Man Can VIII

Bredemann G (1942) Züchtung auf hohen Fasergehalt bei Hanf (*Cannabis sativa* L.). Züchter 14:201–213

Bredemann G, Garber K, Huhnke W, Sengbusch v R (1961) Die Züchtung von monözischen und diözischen, faserertragreichen Hanfsorten (Fibrimon und Fibrida). Z Pflanzenzücht 46:235–245

Christiansen MN, Lewis CF (eds) (1982) Breeding plants for less favorable environments. Wiley & Sons, New York

Clark RB (1983) Plant genotype differences in the uptake, translocation, accumulation, and use of mineral elements required for plant growth. Plant Soil 72:175–196

Clarke JM, McCraig TN (1982) Evaluation of techniques for screening for drought resistance in wheat. Crop Sci 22:503–506

Clarke MF, Adams AN (1977) Characteristics of the microplate method of enzyme-linked immunosorbent assay for the detection of plant viruses. J Gen Virol 34:475–483

Coombs J, Hall DO, Chartier P (1983) Plants as solar collector. Solar energy R&D in the European Community, Ser E, vol 4. Reichel, Dordrecht

Dambroth M, El Bassam N (1983) Low input varieties: definition, ecological requirements and selection. Plant Soil 72:365–377

Earle FG, Melvin EH, Mason CH, Etten van CH, Wolff IA, Jones Q (1959) Search for new industrial oils. I. Selected oils from 24 plant families. J Am Oil Chem Soc 36:304–307

Evans LT (ed) (1975) Crop physiology – some case histories. Cambridge Univ Press, London

Gifford RM, Evans LT (1981) Photosynthesis, carbon partitioning, and yield. Ann Rev Plant Physiol 32:485–509

Graham RD (1984) Breeding for nutritional characteristics in cereals. In: Tinker PB, Läuchli A (eds) Advances in plant nutrition. Praeger, New York, pp 57–102

Hay RKM, Walker AJ (1989) An introduction to the physiology of crop yield. Logman Scientific and Technical, Harlow

Hendriksen AJT, Sneep J (1979) Seed production and distribution. In: Sneep J, Hendriksen AJT (eds) Plant breeding perspectives. PUDOC, Wageningen, pp 370–395

Herzog H (1986) Source and sink during the reproductive period of wheat. Parey, Berlin

Hirsinger F (1980) *Cuphea*: die erste annuelle Pflanze für die Erzeugung von mittelkettigen Triglyceriden (MCT). Fette, Seifen, Anstrichmittel 10:385

Hondelmann W (1987a) Variation, heritability and response to selection in natural populations of caper spurge (*Euphorbia lathyris* L.). Plant Breed 98:65–67

Hondelmann W (1987b) The domestication of caper spurge (*Euphorbia lathyris* L.) – a potential oil crop. In: Eucarpia (Section Genet Resources) Proc of a workshop on Evaluation of genetic resources for industrial purposes. FAL Braunschweig, pp 97–107

Hurd RG, Biscoe PV, Dennis C (eds) (1980) Opportunities for increasing crop yields. Pitman, Boston
International Rice Research Institute (IRRI) (1987) Annual Report for 1986. Los Baños, Philippines
Larsson S, Górny AG (1988) Grain yield and drought resistance indices of oat cultivars in field rain shelter and laboratory experiments. J Agron Sci 161:277–286
Lessman KJ, Anderson WP (1981) Crambe. In: Pryde EH, Princen LH, Mukherjee (eds) New sources of fats and oils. Am Oil Chem Soc, Champaign, pp 223–246
Lind V (1983) A test system for the quantitative assessment of resistance to mildew in rye (*Secale cereale* L.). Phytopathol Z 108:127–134
Loughman BC, Roberts SC, Goodwin-Bailey CI (1983) Varietal differences in physiological and biochemical responses to changes in the ionic environment. In: Saric MR, Loughman BC (eds) Genetic aspects of plant nutrition. Nijhoff, Den Haag, pp 175–187
Mares DJ (ed) (1987) Fourth international symposium on pre-harvest sprouting in cereals. Westview Press, Boulder, Colorado
Müntz K (1987) Symbiotic nitrogen fixation of legumes: discovered 100 years ago – what do we know at present? Biol Zentrbl 106:547–567
Mussel H, Staples RS (eds) (1979) Stress physiology in crop plants. Wiley & Sons, New York
Plank JE vd (1982) Host-pathogen interaction in plant disease. Academic Press, New York
Quizenberry JE (1982) Breeding for drought resistance and plant water use efficiency. In: Christiansen MN, Lewis CF (eds) Breeding plants for less favorable environments. Wiley & Sons, New York, pp 193–212
Raymond CA, Harwood CE, Owen JV (1986) A conductive method for screening populations of eucalypts for frost damage and frost tolerance. Aust J Bot 34:377–393
Röbbelen G (1982) Möglichkeiten züchterischer Veränderungen von Kultur- und Wildpflanzen zur Erzeugung technischer Öle und Fette. In: Fischbeck G, Haushofer H, Schroeder D, Wiebecke C (eds) Agrar Spectrum Schriftenreihe 4. Verlagsunion Agrar, pp 369–390
Rohrmoser K, Seidel H (1977) Gesunde Saat gute Erträge. Deutsche Ges.f Techn Zus.arbeit (GTZ) Eschborn
Saastamoinen M (1986) The possibilities to increase biological nitrogen fixation by breeding. Ann Agric Fenn 25:147–155
Turner NC (1986) Crop water deficits: a decade of progress. Adv Agron 39:1–51
Umaerus V (1969) Studies on field resistance to *Phytophthora infestans*. 2. A method of screening young potato seedlings for field resistance to *P. infestans*. Z Pflanzenzücht 61:167–194
Vavilov NI (1950) Selected writings of NI Vavilov: the origin, variation, immunity and breeding of cultivated plants. Chron Bot 13
Vose PB, Blixt SG (eds) (1984) Crop Breeding – a contemporary basis. Pergamon Press, Oxford
Wallace DH (1985) Physiological genetics of plant maturity, adaptation, and yield. Plant Breed Rev 3:21–167
Werner D (1987) Pflanzliche und mikrobielle Symbiosen. Thieme, Stuttgart
Wezenbeek P v, Vos P, v Boom J, v Kammen A (1982) Molecular cloning and characterization of a complete DNA copy of potato spindle tuber viroid RNA. Nucl Acids Res 10:7947–7957

5 The Plant Breeding Station

The best breeding methods will not be successful if the technical prerequisites for extensive selections and comparative performance trials are not present in the individual steps of the breeding process. For the evaluation of breeding success, appropriate bookkeeping is an important resource, the data of which can also simultaneously provide indications for further planning. After a new variety has been completed, the further task for the breeder is its propagation, which is connected at the same time with the setup of maintenance breeding. Since it cannot be the task of this compendium to describe the exceedingly manifold working operations of a breeding station and the necessary equipment, only some indications will be given.

5.1 Tools for Plant Breeding

The higher the standard reached by breeding in a country, the greater the effort will necessarily be for further breeding work and the smaller the progress still to be achieved.

5.1.1 Mechanization

A rationalization and mechanization of all breeding processes have taken place in all countries with highly developed plant breeding for the following reasons: to save labor, to decrease costs, and to accelerate working operations. An interruption in the establishment and harvest of the trials due to periods of bad weather has an unfavorable effect on their interpretation. In past years, special machines and equipment for all operations in breeding have been developed in many places, also with regard to the special requirements of different cultivated plants. They include lay-out of the nursery, preparation for sowing, also sowing of the smallest quantities of seed with precisely functioning and easily cleaned sowing machines, and in addition, employment of machines for the uniform distribution of fertilizer and the fully mechanized harvest of trial plots, etc.

5.1.2 Trial Design

The choice of trial methods, the results of which influence whether a family or a breeding strain will be include in or eliminated from the further breeding process, is determined by the number of test lines and the available quantity of seed. In cereal breeding, the number of A lines to be tested often amounts to several thousand. The seed quantity for each A line originating from one plant or occasionally even from only one single ear is so small that in the sowing of single seeds the plots are no larger than 0.5–2.0 m^2 (two to four rows of 1–2 m each). The breeder is faced with the difficult task of evaluating a large number of small plots, which are set up without replications. Another difficulty in the evaluation arises from the fact that, along with the number of lines, the size of the required area also increases.

Quite a large number of trial designs are suitable for evaluation of breeding lines, such as randomized block, strip, split-plot, and lattice designs. Lattice designs are particularly suitable for tests since the blocks for this kind of design need not be complete. They permit the inclusion of a very large number of treatments, i.e., test lines or families. In order to achieve the same effectiveness as with lattices, randomized block designs generally require more replications. Regarding the distribution of a new breed over wide areas, it is necessary to also carry out trials at other locations as early as possible, to realize the highest possible turnover and profit to cover breeding costs. Trials with less replications at several locations generally provide more reliable results than a single trial with numerous replications. Moreover, replications of trials in different years furnish additional valuable information.

One quantitative measure for the ecological adaptation of the potential cultivars is ecovalence, a breeding parameter developed by Wricke (1965). Such trial evaluations permit ascertainment of areas for which breeding of special varieties may be appropriate.

5.1.3 Evaluation and Documentation

For the evaluation of extensive performance trials, the use of computers is a possibility. Since cheap personal computers (PCs) with enormous capability have been put on the market in recent years, it is now also possible for the smaller breeding firm to quickly process the data obtained in trials, using either self-developed or purchased software. Such small computers are able to quickly furnish the breeder with a maximum of information having a high degree of reliability as to the behavior of his breeding lines and his breeding progress, just like the larger computers, which are still needed for the evaluation of extensive quantities of data.

The problem of the evaluation of breeding material is closely related to the necessity of keeping appropriate breeding records. Their purpose is to simplify sowing, observations, harvest, and processing of the trials, furthermore, to provide information on the origin and size of the breeding material, and to permit conclusions not only for the evaluation of the respective results but also for planning and carrying out new trials. Considering the fact that today breeding projects last for relatively long periods of time of ca. 10–20 years and must therefore be planned both long-range and at the same time flexibly, and that often several and occasionally even alternating skilled workers are involved in their execution, the necessity of an objective and easily accessible, as well as clearly arranged, data collection is evident. Opinions differ as to which is the most appropriate was to keep records; they are based on the different objectives and aims, the given work possibilities and, finally, the personal experience and preferences of the breeder.

Recently, the use of electronic devices has had an increasing importance in plant breeding, too. The necessary data can also be directly entered into the computer, for example, from a scale, whereby the risk of clerical errors leading to wrong results is considerably reduced. Moreover, there are electronic data processing devices available today for evaluations in the field. These are small, handy computers, which can be programmed before beginning the evaluation, so that plot numbers appear in the same order as they were arranged in the nursery. Hence, walking from plot to plot, the evaluator can enter his evaluation reports using a keyboard or with the help of an electronic scanner. After completion of the work, the data recorded in the hand-device can be transferred to the stationary computer, which records these data on suitable disks. It does not require a great effort to print these data in a suitable form on a list, which should be done in any case for control and security. Nowadays, efficient, handy, and weatherproof "laptops" are available, which can be taken along into the nursery. Suitable software for breeding firms that may helpfully support the realization of long-standing breeding programs are generally available today.

This trend of modern collection and processing of data, briefly described here, will be most important in the future because of the growing concentration of breeding enterprises, the increasing scope and duration of breeding projects, and the necessary efficiency of all work processes. Finally, it is once more emphasized that each system of data collection is, in terms of its utilization, only a resource – if a very important one. It does not exempt the breeder from the task of familiarizing himself, by repeated critical observations, with the different behavior of his breeding lines and the respective environmental conditions of the individual trial varieties, such as weather conditions, soil conditions, progress of epidemics, effect of plant protection measures, et al. This requirement should not be misunderstood to the point that prime importance for breeding success is attributed to the so-called breeder's green thumb.

5.2 Classical Maintenance Breeding

The breeder, in his own interest, must take care that the character and hence the genotypic composition of his varieties are preserved. The necessary measures to be taken for this are comprised within the term maintenance breeding. Some important aspects will now be explained.

In self-fertilizing crops, the breeder will generally be satisfied with the cultivation of a more or less large number of A lines, deriving from selection of single plants from the A lines of the previous year or from a multiplication stage of a breeding cycle. The A lines are only phenotypically evaluated and, after elimination of on the average 10%, the remaining ones are propagated together as first "base seed" or "elites." How extensive these elites and thus the number of A lines to be tested must be depends on the seed demand of the market and the multiplication quota of the variety. Here, it must be considered that not more than five to six multiplication steps should be inserted between base seed and seed for sale.

In varieties originating from selection within an early cross generation, perhaps, F_3 or F_4, a more intensive maintenance breeding may initially be necessary, the extent of which is determined by the degree of the remaining heterozygosity of the generation for selection. For this, the best A lines of a B and C line trial are tested and the unusable lines are subsequently eliminated, so that the first base seed is produced by multiplying the best C lines together. With this simple method of continuous selection of individuals followed by progeny tests, the required uniformity in the characters determined by description of the variety can be attained.

With this somewhat more intensive maintenance breeding, the breeder runs the risk that the variety is also changed in a positive direction, so that it does not conform to the required variety characterization any longer and hence looses it license. The variety so improved must first be subjected to a long and costly trial procedure to acquire a license as a new variety, before it can again be used commercially by the breeder.

In cross-pollinated crops, intensively pursued maintenance breeding is generally necessary to counteract the selection pressure through natural selection on the recombinations occurring within each generation. Frequently, maintenance breeding requires just as much effort as breeding a new variety, especially in rye and sugar beet varieties, which are widely distributed. By the application of intensive selection methods with regulation of cross-pollination and testing for combining ability of the components of the variety, described in Sections 2.1.3 and 2.4, continuous improvements of the varieties can also be achieved. However, these are so small in the consecutive breeding cycles, especially when they are quantitative changes in characters determining yield, that they cannot be recognized by the registering authorities. In addition to this, the admissible latitude concerning variability of variety characters is larger in cross-pollinated than in self-fertilizing crops. Greater progress in the performance of a variety can only be recognized if years of base seed production, widely divergent chronologically, are compared.

In varieties that are less widely distributed and for which the effort of intensive breeding methods is not economically feasible, the breeder should restrict himself to mass selection. It is more effective than individual selection with progeny tests carried out on a narrow basis. If a breeder must work with several varieties of a species at the same time, as is often the case in vegetable breeding with cabbage, carrots, beet roots, and others, it is recommended that enough seed be produced every year so that starting seed for multiplication is available for several years. Its ability to germinate can be maintained by appropriate storage conditions (low temperatures, low atmospheric humidity, vacuum containers and, circumstances permitting, also freeze-drying). In the multiplication of cross-pollinated varieties, environmental conditions must especially be considered. In order to achieve high seed harvest and seed quality, fodder plants and vegetable species are preferably multiplied in other countries that have more favorable climatic conditions. Under these circumstances, it is particularly important that maintenance breeding and the first multiplication steps take place in the country of seed demand and that only the last stage of the breeding cycle is multiplied abroad.

Maintenance breeding in vegetatively propagated plants concentrates above all on the production of plant material free of disease, which is transferred to the following generations by reproductive organs. In potato, the methods of early diagnosis proved to be a success. By raising buds in greenhouses during the winter months or by a serological test, large quantities of individual tubers can be assayed for virus diseases. This pretest is followed by an extensive testing of A clones and possibly even of B and C clones, from which stock plant material is produced. If further multiplication steps take place, mass selection must be carried out by removing plants with virus diseases as infection sources from the field.

5.3 New Forms of Maintenance Breeding

In the field of maintaining an existing cultivar protected by breeders' rights, biotechnological methods have also become integral parts of applied breeding programs. Horticulture and the maintenance of vegetatively propagated crops such as the potato rely on tissue culture techniques that allow for the rapid propagation of specific clones in vitro. The first plants mass-propagated by tissue culture were ornamentals, and they lie at the root of the first commercial tissue culture facilities. Vasil and Vasil (1980) published a table of 352 species in which adventitious buds, shoots, embryoids, or whole plants were produced aseptically in vitro on synthetic media. Again, the Murashige and Skoog medium (Table 16) is the most universal plant propagation medium. Normally, the explant forms callus first, which can be induced by external treatment with phytohormones, auxins, and cytokinins to regenerate shoots and roots. Depending on the size of the explant, this technique is called rapid propagation (up to 1-cm ϕ large pieces), or meristem culture (pieces normally around 0.2 ϕ mm).

5.3.1 Rapid Propagation

Rapid propagation gained economic importance in a number of ornamentals such as carnation, gerbera, dahlia, orchids, and *Pelargonium*, but also in strawberries, grapes, fruit and forest trees, palms, asparagus and, within the field crops, in potato. Potato breeders, for example, use per company per year 50 000-100 000 rapidly propagated plants, which are transferred from the in vitro culture to the greenhouse in early spring. Tubers from such plants can be harvested in May before aphids start to fly and transfer viral diseases. This use of rapid propagation has opened new practical ways for cheaper maintenance breeding, and is being used increasingly by commercial companies.

This technique of rapid propagation is easier to apply in dicots than in monocots, as monocots cannot be regenerated from leaf or shoot pieces. Therefore, in monocots, cultures are started from immature embryos between 0.5 and 1.5 mm. In vitro, the scutellum will form callus that is either regenerated directly or taken into suspension culture. Although substantial progress has been made in the regeneration, maintenance, and propagation of monocots starting from embryogenic suspension cultures, e.g., in *Pennisetum americanum, Panicum maximum, Zea mays,* triticale, *Hordeum vulgare, Triticum aestivum,* and *Oryza perennis*, it should be stated that rapid propagation in field crops is so labor-intensive that it is only economical in dicots.

5.3.2 Meristem Culture

Rapid propagation techniques suffer because the total offspring from a diseased plant will be diseased as well, particularly by virus infections, which cannot be controlled chemically and can cause severe problems. Consequently, plant material for rapid propagation must first be taken through meristem culture. Normally, up to 40% of regenerants from apical meristems smaller than 2 mm will deliver healthy plants. By culturing the starting material at 30–35 °C, the percentage of healthy plants can be increased. However, the optimal size for the meristem should always be tested first. The larger they are, the easier the procedure is, but viruses may be transferred. In a comparative study with different potato viruses, it was demonstrated that only for potato virus X was the size critical, while for the purification from potato virus S, even 2-mm tips could give virus-free regenerants. These larger tips regenerate much more early and never produce callus, while smaller ones tend to proliferate in an unorganized manner. Since a given genotype must remain stable, it is important to prohibit this callus formation.

5.3.3 Artificial Seeds

The micropropagation system would be further advanced if the laborious transfer from in vitro to in vivo could be avoided. Nature uses seeds with an embryo embedded in the endosperm for this purpose. The production of embryoids and embryogenic suspension cultures has become possible for a number of species. In the late 1970s, the idea arose to coat such embryos with an artificial nutritional medium and to produce an artificial seed. The first prerequisite for this approach is a really good embryogenic suspension, such as that available in carrots since 1958 or in rapeseed. The quality and fidelity of induced embryo genes are the limiting factors for the development and scale-up of artificial seeds. According to Redenbaugh et al. (1987), artificial seeds will most likely be used first for hybrid vegetable crops such as celery, *Brassica* species, carrot, and tomato but also for cotton, lettuce, and alfalfa. Water-soluble hydrogels are suitable for making artificial seeds. Redenbaugh et al. (1987) mixed alfalfa somatic embryos with 2% sodium alginate suspensions and dropped them into 100 m*M* $Ca(NO_3)_2$. Surface complexation began immediately and the drops were gelled. For applied purposes, the costs are still much higher than for normal seed (Redenbaugh et al. calculated US $0.026 for one capsule), but for specific purposes, e.g., for transformed material, the procedure may be economical.

5.3.4 Long-Term Storage

Particularly for vegetatively propagated and/or polyploid crops, the maintenance of pathogen-tested stocks is always a problem. Here, living collections are transferred on modified media just once a year and stored at temperatures between 4 and 10 °C. Storage in the form of tissue culture has been proposed as an alternative to traditional clonal methods for the conservation of potato germ plasm. The growth limitation method compared to the traditional clonal storage offers a real savings in cost, time, and effort. After culture of about 4 weeks at 20 °C, the plantlets are transferred at a size of 3–4 mm to the long-term storage conditions at 10 °C. The duration of the culture in liquid medium lasts for about 2 years.

In order to reduce the chance of genetic change occurring even during storage in living collections, the use of cryopreservation (preservation in the frozen state) has been proposed by Withers and Street (1977). Cryopreservation is based on the reduction and subsequent arrest of metabolic functions of plant material by imposition of an ultra-low temperature of –196 °C. However, only a few biological materials can be frozen in their natural state without adverse effects on cell viability. The discovery of chemicals with cryoprotective properties was a prerequisite for this method, in particular glycerol and dimethyl sulfoxide (DMSO). The ideal material to start with from the practical point of view would be meristems. Unfortunately, this strategy can only be employed with about ten species at the present time, because of the difficulty that is encountered in devising suitable cryopreservation procedures for meristems. For potato meristems an improved two-stage procedure exists: an initial, slow-cooling stage (0.3 °C min^{-1}) followed by immersion in liquid nitrogen for the second, rapid-cooling stage. Ten percent DMSO is used as a cryoprotectant. Today, more than 40 plant species are cryopreserved as cell, callus, protoplast, or meristem culture.

5.4 Plant Breeders' Rights, Variety Listing, and Seed Control Systems

Seed of clearly defined varieties is of great importance in improving the yield and quality of agricultural and horticultural products. It is for this reason that many states devote considerable effort to promote the breeding of new varieties and to stimulate the production of first-class seed. Already 30 years ago the importance of this was underlined by the Food and Agricultural Organization of the United Nations (FAO) in declaring 1961 as World Seed Year with many specific activities in a great number of countries. In that year many initiatives were started for long-term development.

One of the most important incentives to encourage plant breeding is the grant of industrial property rights for new varieties. Already in 1833 a Papal edict allowed such a possibility for Vatican City. Later in the last century, patent legislation was introduced in a number of European countries and in North America. But soon it was felt that the patent right as such was not appropriate to living plant material as plant varieties.

Therefore, a specific form of industrial property right was adapted to include plant breeding, plant varieties, and their reproductive or vegetative propagating material. This was done through the International Convention for the Protection of New Varieties of Plants, which was ratified in Paris at the end of the World Seed Year in December, 1961, and which took effect in 1968. Only with the help of an industrial property right is the breeder certain to obtain regular renumeration for the enormous work connected with the creation of a new valuable variety. The most substantive provisions of the Convention follow.

- A protection right can be granted for varieties of all botanical genera and species.
- The result of the right granted to the breeder or discoverer of a new variety is that he alone is authorized to commercially produce or market propagation material.
- The authorization of the holder of the right is not necessary if the new variety is used as basic material for the creation of further new varieties. This does not apply, e.g., to the use of protected inbred lines for the production of hybrids.
- The new variety must be clearly distinguishable, sufficiently homogeneous, and stable in its essential characteristics.
- The new variety must have a designation that adheres closely to the rules of the International Code of Nomenclature.
- The minimum period of protection is 15 years. For vines, forest trees, fruit trees, ornamental trees and shrubs including, in each case, their rootstocks, the period of protection is not less than 18 years.
- States abiding by the Convention are entitled to extend the right for varieties of certain genera or species, under specific conditions, to the final product (i.e., cut flowers or fruits).

All states party to the Convention constitute a union, called the Union for the Protection of New Plant Varieties (UPOV), which has its seat in Geneva. At present, 18 states are members of the UPOV (Table 32). These states work closely together in technical, administrative, and legal matters. In 1988 within the UPOV member states, applications to obtain plant breeders' rights were submitted for ca. 5000 new varieties. Many other countries worldwide that already grant

Table 32. Member states of UPOV (1989)

Australia	Japan
Belgium	The Netherlands
Denmark	New Zealand
France	South Africa
Germany	Spain
Hungary	Sweden
Ireland	Switzerland
Israel	United Kingdom
Italy	United States of America

industrial property rights to plant varieties or that intend to implement them are interested in the work of UPOV or even participate in the practical work of that organization. Such countries are Austria, Canada, Czechoslovakia, Mexico, Morocco, Norway, Poland, and others.

In cooperation with international plant breeders' organizations, UPOV has introduced guidelines that can be applied in growth tests in the field and in greenhouses or laboratories, to determine distinctness, homogeneity, and stability. The basis of the guidelines are the characteristics of plants and their organs for varieties of different species. These are not necessarily qualities that characterize a certain value of the variety. They must be capable of precise recognition and description. The guidelines prescribe, species by species, the size of plots, the number of plants per plot, the number of plants or organs for which characteristics must be assessed, the number of replications, and the duration of the tests. Finally, they explain the statistical principles for the assessment of distinctness, homogeneity, and stability.

These guidelines have the advantage of being internationally comprehensible and therefore affording clarity on what is understood by a variety from a technical and juridical point of view. They should never be used restrictively in such a way that progress in plant breeding would be hampered. Therefore, they must be constantly and rationally adapted to changing circumstances. On the basis of the guidelines, the member states of the UPOV have divided their work so that variety testing is, in many cases, concentrated in such a way that certain member states do the practical testing for certain others.

Besides granting plant breeders' rights, many states worldwide are involved in other activities with respect to plant varieties and seeds. They have introduced seed laws or at least rules to ensure that, for agricultural crops and vegetables, only seed of such varieties is marketed that is suited to the country in question. After semi-official or official growth tests, only those varieties that meet the standards are accepted for a national variety list.

In most cases, the requirements for distinctness, homogeneity, and stability of the varieties are the same or approximately the same as were explained for plant breeders' rights. However, for variety listing the main topic is the testing for value and use. For 2 or 3 years, the candidate varieties are tested in trials

throughout the whole country or at least over a certain region. The variety trials are normally combined with agronomic tests in such a way that the amounts of fertilizers and pesticides are altered; the trials include resistance tests and tests with respect to quality (such as baking or malting quality in wheat or barley). After a statistical evaluation of the tests, new varieties are listed and recommended for agriculture and horticulture. In West European countries such as the Federal Republic of Germany, France, or the United Kingdom, each year up to 1000 varieties are applied for registry into the national variety lists.

In addition to this in many states, independent of their state of development or their social statutes, seed production and distribution are regulated to the extent that in agricultural crops seed of the category "Certified Seed" is only allowed to be produced from Basic Seed, and this only from Pre-Basic or Breeders' Seed. Very often, corresponding rules apply to vegetable seed. The seed produced must meet specific minimum requirements, such as trueness to variety, germinative capacity, and technical purity. Seed containers must be sealed and properly labeled.

Variety listing and seed control systems exist not only on a national basis. In the early 1950s, the FAO was concerned with the creation of certification rules for hybrid maize. In the late 1950s and 1960s, the Organization for Economic Cooperation and Development (OECD), with its seat in Paris, took over this task. The OECD created certification schemes for cereals, sugar beet, fodder plants, oil and fiber crops, and vegetables. These schemes have been adopted by up to 40 countries (Table 33).

Table 33. Participating countries in the OECD scheme for herbage seed (1989)

Australia	Kenya
Austria	Luxembourg
Argentina	Morocco
Belgium	The Netherlands
Bulgaria	New Zealand
Canada	Norway
Chile	Poland
Cyprus	Portugal
Czechoslovakia	Romania
Denmark	Spain
France	South Africa
Finland	Sweden
Germany	Switzerland
Greece	Tunisia
Hungary	Turkey
Iceland	United Kingdom
Ireland	United States of America
Israel	Uruguay
Italy	Yugoslavia
Japan	

The objective of the OECD Seed Schemes is to encourage the use of high quality seed of varieties and to facilitate the movement of such seed in international trade. The schemes authorize the use of labels and certificates for certain categories of seed produced and processed for international trade (Fig. 63). The varieties of the seed must be listed nationally.

Another international seed certification system is that of the European Economic Community (EEC). The principles of these standards for the listing of varieties and the production and commercialization of seed are similar to those of the OECD.

All of these national and international seed control systems ensure that the varieties in question are maintained in a proper way. This is the only possibility of guaranteeing that the consumer really obtains seed of the variety he wants.

One of the most often used tools for the control of proper maintenance of a variety are the so-called pre- and postcontrol tests. The maintenance control can be done either by the breeder himself or his authorized agent, or by designated authorities who are responsible for granting plant breeders' rights, variety listing, and/or seed certification. In these tests, comparisons are always made between a reference or standard sample of the variety and samples that are drawn from Pre-Basic Seed, Basic Seed, or Certified Seed. It is not necessary in all cases to continue the control tests throughout the whole vegetation period; sometimes, as is often the case in vegetable control tests, it is possible to make the comparison based on young plants in the greenhouse. Some of the other modern techniques in maintenance control are the different methods of electrophoresis.

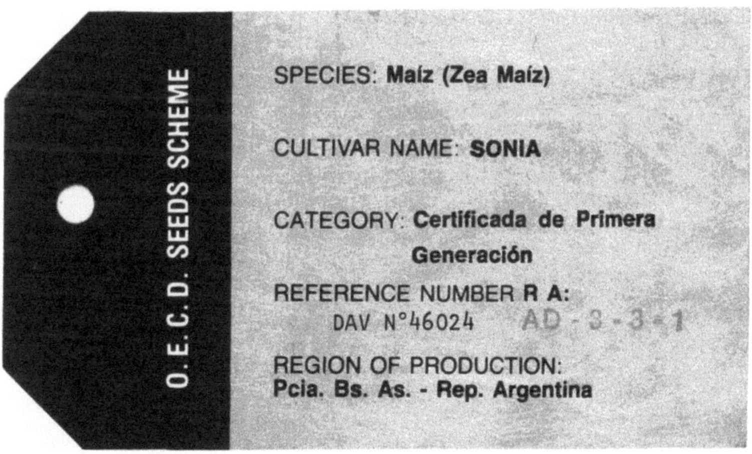

Fig. 63. The OECD label

5.5 Cited and Recommended Literature for Chapter 5

Böringer D (1978) Technical and legal aspects of the protection of new plant varieties and the use of plant breeders' rights in the marketing of seed. Subregional Andean Forum on Seed, Lima
Böringer D (1980) Examination of new varieties of plants for distinctness, homogeneity and stability. 9th Pan-American Seed Seminar, Buenos Aires
Böringer D (1988) Industrial property rights and biotechnology. Gazette and Newsletter of UPOV 55, Geneva
Catalogue commun des variétés des espéces agricoles – quatorzième edition intégrale (1987) Journal officiel des Communautés européennes (87/C 352 A/01) Bruxelles
Cochran WG, Cox GM (1957) Experimental designs. 2nd edn. Wiley & Sons, New York
European Economic Community (1966) Council directive on the marketing of cereal seed (66/402/EEC), Brussels
European Economic Community (1970) Council directive on the common catalogue of varieties of agricultural plant species (70/457/EEC), Brussels
Falconer DS (1981) Introduction to quantitative genetics. Longman, London
FAO (1975) Cereal seed technology. Rome
Gomez KA, Gomez AA (1984) Statistical procedures for agricultural experiments, 2nd ed. Wiley & Sons, New York
ISTA (1967) Seed legislation. vol 32, No 2
International code of nomenclature for cultivated plants (1980) Bohn, Scheltema + Holkema, Utrecht. Junk, The Hague
International union for the protection of new varieties of plants (1987) The first twenty-five years of the international convention for the protection of new varieties of plants. UPOV, Geneva
Jungenheimer RV (1976) Corn, improvement, seed production and uses. Wiley & Sons, New York
Kartha KK (1987) Advances in the cryopreservation technology of plant cells and organs. In: Green CE et al. (eds) Plant tissue and cell culture. AR Liss, New York, pp 447–458
Kunhardt H (1989) Industrial property rights and their impact on industry and agriculture. Proc XII Eucarpia Congr, Göttingen, Parey, Berlin, pp 293–306
OECD Scheme for the varietal certification of herbage and oil seed moving in international trade (1987). Organization for Economic Co-operation and Development, Paris
Redenbaugh K, Viss P, Slade D, Fujii JA (1987) Scale-up: artificial seeds. In: Green CE et al. (eds) Plant tissue and cell culture. AR Liss, New York, pp 473–493
Steel RGD, Torrie JH (1982) Principles and procedures of statistics, 3rd printing. McGraw-Hill, Auckland
Thomson JR (1979) An introduction to seed technology. Leonhard Hill, Glasgow
Union international pour la protection des obtensions végétales (1972) Actes des conférence internationales pour la protection des obtentions végétales: Publication UPOV 316, Genève
Vasil IK, Vasil V (1980) Clonal propagation. Intern Review Cytol Suppl 11A:145–173
Withers L, Street HE (1977) The freeze-preservation of plant cell cultures. In: Barz W et al. (eds) Plant tissue culture and its Bio-technological Application. Springer, Berlin Heidelberg New York pp 226–244
Wricke G (1965) Erfassung der Wechselwirkung zwischen Genotyp und Umwelt bei quantitativen Eigenschaften. Z Pflanzenzücht 53:266–343

6 Safeguarding and Utilization of Natural Genetic Diversity

The decline of variation in crop plants is traceable to diverse causes. In the discussion of breeding methods the drastic influence of selection became evident. By this means the large variation of first generations after crossing decreases and will finally vanish. Also, the more successfully a newly bred cultivar expands its growing area, the less other cultivars can keep their positions. Moreover, leading cultivars very often are closely related, as can be traced from numerous pedigrees. With cross-pollinating species the change from open-pollinated varieties to hybrid varieties, as frequently can be observed, considerably contributes to the narrowing of the genetic base, which at the same time means abandoning the heterogeneous varietal type in favor of the homogeneous and more uniform type.

6.1 Genetic Vulnerability of Crop Plants

The uniformity of cultivars, which plant breeders strive for because of market demands, can produce catastrophic effects. This was the key lesson of the southern corn blight epidemic in 1970 in the USA. There, since the late 1950s, the Texas cytoplasm was more and more made use of for the production of maize hybrids, due to its great environmental stability. By 1968 it had been introduced into 75–90% of commercial maize hybrids. The pathogen of corn blight [*Cochliobulus heterostrophus*, syn. *Helminthosporium maydis*] which, although known, was of limited importance, apparently found with this cytoplasm a suitable background for mass propagation. In 1970 an epidemic-like outbreak of the disease appeared, which resulted in a yield loss of 50% in some southern states and 15% nationwide. Since then the genetic vulnerability of crop plants favored by uniform varieties is a well-established fact. Also, other major crops in the USA exhibit considerable uniformity. In comparison, the situation in Europe is not as threatening, even though some bottlenecks exist. Bosemark (1979) called for broader cytoplasmic diversity for sugar beets and for a broadening of nuclear gene variation to insure continuing breeding progress.

Historically, there have been other plants epidemics that can be traced back to the genetic uniformity of species. The blight disease of potatoes in Ireland in 1846, which resulted in widespread starvation, is probably the most well known of all epidemics. Furthermore, the Panama banana disease at the beginning of this century and the *Phylloxa* disease in French grapes caused by root aphids around 1850 should also be mentioned. The cultivars of these crop species are clones and are, therefore, genetically uniform. With all of these crops, the objective was to minimize uniformity by introducing resistant cross parents or, in the case of maize, by incorporating alien cytoplasms for the production of male-sterile cross parents, i.e., in general to broaden their genetic diversity. The parental diversity or genetic divergence of cross parents must be complemented by regional diversity, i.e., by genetic differences within and between crop plantings in one region.

For the first time in years, it can now be confirmed that the increase in uniformity has been halted for some crop species or even reversed. Duvick (1984) made the general statement for the USA that, at the beginning of the 1980s, the genetic basis of field crops was not as narrow as it was in 1970. In the red winter wheat regions of the USA, for example, an increase of diversity as a consequence of introducing new gene pools could be noted, both for hard and soft red winter wheat cultivars.

6.2 Landraces and Wild Species

The question now is: which plant taxa, in contrast to modern elite cultivars offer genetic diversity? Genetic diversity is above all a typical characteristic of landraces, which are actually complex evolutionary populations. These landraces were not only the progenitors of recent cultivars, but also of their initial germ plasm. In addition, the wild relatives of crop species provide large genetic potential.

Nowadays, landraces can be found mainly in ecological niches, i.e., within close boundaries of more or less extreme environmental conditions. Mountainous sites provide an example of these types of conditions. On the Austrian slopes, existing varieties of wheat exhibit phenotypic uniformity with respect to morphological traits. This is the result of continuous mass selection of the phenotype. By means of natural selection at the locations involved, genetic complexes of adaptability to changing climatic conditions have been established over the years. A range of variation of physiological characters has also been maintained. Yields are certainly low, but yield stability is very high and field resistance for parasitic diseases is striking. Specific genes for specific pathotypes of parasites are not present, but an unspecific resistance controlled by polygenes does exist. Of special interest for breeding is the occurrence of genotypes with high protein content and high-grade amino acids and a strongly expressed competitive trait against weeds, so that the use of herbicides becomes unnecessary.

The removal of landraces by phenotypically and genetically uniform elite cultivars had the temporary effect of overlooking their breeding value. Recently, however, it has been recognized that landraces look promising even in advanced breeding programs. In view of the increasing interest in still existing landraces or those maintained in working collections of plant breeders, new collecting activities for supplementing this varietal spectrum not only include major crop plants such as cereals, roots, and tuber crops, but also minor ones such as vegetables and fodder crops. The results obtained so far are very encouraging.

In particular, the collecting expeditions of the "Institut für Genetik and Kulturpflanzenforschung in Gatersleben" (Germany), which were organized cooperatively with national institutions, revealed a large variability between and within species. Even in more industrialized countries with a highly developed agriculture, surprisingly many old varieties and genotypes could be found, especially autochthonous species in house gardens, which are commonly considered typical examples of a less developed agriculture. The Plant Genetic Resources Newsletter of IBPGR contains worldwide information on such collecting missions.

Besides landraces, other genetic resources are available to the plant breeder. Of the various categories, the wild species must be emphasized. They are of increasing importance, especially for transferring resistance but also as a source of physiological and morphological traits. In winter barley, for example, the possibility cannot be excluded that a combination of the extremely long vernalization period of wild barley (*Hordeum spontaneum* C. Koch) with the best frost hardiness of cultivated forms could result in an improvement of the often inadequate hardiness of winter barley (Schwarzbach 1986). In addition, Schwarzbach points out the extreme hairiness of some wild forms, the function of which is widely unknown, but which could serve as protection against insects or to catch pathogenic spores. Although after crossing wild germ plasm with cultivated forms, many unwanted wild-type characters must be eliminated, the utilization of wild germ plasm, i.e., *H. spontaneum*, is considered essential.

An analogous situation exists with a number of other crop species. Some wild species of the potato have become well known. A list of European potato cultivars published in 1986 contains 97 cultivars carrying genes from wild species. Altogether, 11 wild species have been introduced in European potato breeding (Ross 1986). Besides their utilization for resistance breeding, the importance of *Solanum* wild species for heterotic effects is obvious. Another inventory shows that in wheat over the course of 5 years, from 1980 to 1984, about 10% of all crosses were with landraces and foreign varieties, wild species of *Triticum* and *Aegilops* as well as of more distant relatives, sometimes as part of a complex pedigree. Disease and pest resistance along with stress tolerance were the outstanding characters that were utilized.

While the above refers to major crop plants, it has to be pointed out that in minor crop plants there is a deficiency in utilizing wild germ plasm. There seems to be a rather strong correlation between economic importance and the economic possibilities for plant breeders. On the other hand, one should not overlook the fact that the direct use of wild germ plasm retards a breeding program by at least several years. Therefore, so-called germ plasm enhancement or pre-breeding is a necessary step in overcoming barriers between wild germ plasm and advanced lines. Moreover, wild species are species in their own right, and can be considered as a resource for new cultivated plants. Especially in the industrialized countries nowadays, new crop plants are used for industrial purposes because of their particular composition (see Section 4.4).

The large number of medicinal plants and spices should also be mentioned. The basic requirement for developing new crops is an evaluation of wild species for pure plant stands, adaptation to field conditions, and crop rotation systems. In addition, investigations on the quantitative and qualitative variability of important plant characteristics must be made, including a representative sample of populations throughout the natural distribution area. Finally, the accumulated knowledge of the evolution of known crop species must be utilized to develop new crops from the wild; examples here are the biological mechanisms that were identified by cytogenetic and taxonomic studies in the evolution of wheat.

Looking at the germ plasm as a whole, which can be exploited as a gene reservoir by plant breeders, a certain diversity is found. This diversity springs from three sources: the centers of diversity, the centers of cultivation, and the breeding programs (Fig. 64). These categories, however, differ to some extent in respect to genetic composition, variation, agronomic value, and genetic potential for breeding. The centers of diversity listed in Fig. 64 are plant genetic resources in the narrow sense, because they form the basis also for the two other resources.

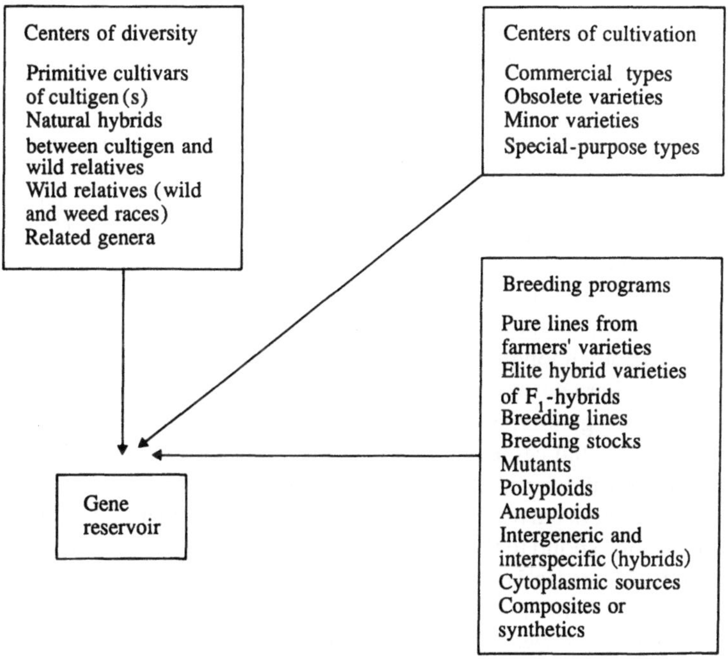

Fig. 64. The spectrum of germ plasm in a crop species (cultigen) and its sources (Sneep and Hendriksen 1979)

6.3 Centers of Diversity

It is to his great credit that the Soviet geneticist, N.I. Vavilov (1887–1943) investigated regions of great diversity in a systematic, global way. By means of extensive expeditions, he explored large areas of the world from 1920-1943, and made an inventory of cultivated plant species, their primitive and wild relatives. After a comprehensive evaluation of collections, he identified a concentrated diversity of certain crop species in some regions of the world. These regions, separated by mountains, prairies, or deserts for the most part, were located in the northern temperate climatic zones between 20 and 45 degrees of latitude. They were designated gene centers or centers of origin of crop plants. Vavilov summarized the results of this broad research work in his theory of gene centers (Vavilov 1928), the impact of which was manifold (cited from Lehman 1984):

1. It condensed the knowledge of the 1920s and 1930s on the origin, descent, and diversity of crop plants, interpreted it, and put it into a logical context.
2. It stimulated in a unique way all branches of crop plant research, from taxonomy and evolution research to plant physiology and biochemistry.
3. It accelerated the transformation of plant breeding from an empirical to a scientific basis. Moreover, it encouraged the establishment of worldwide collections and thorough studies of the huge wealth of genetic resources of our crop species and their relatives.

As a consequence, an increasing number of collecting expeditions were mobilized from Europe alone; more than 20 expeditions were organized between 1925 and 1980 just for potatoes. Thus, it was proven that centers of diversity occurred not only at restricted higher elevations separated by mountains, deserts, or prairies, but also in the lowlands. In the course of an inventory of maize in Mexico, Columbia, and Brazil, it was found that diversity in the tropical lowlands was greater than in mountainous areas, a result of huge plains covering several degrees both of longitude and latitude. Vavilov's theory that the middle of a gene center is characterized by an accumulation of dominant alleles, while an increase of recessive genotypes occurs towards the periphery was invalidated. Even the assumption that center of diversity and center of origin are identical was found not to be valid for all species. For example, centers of diversity of wild barley were detected, particularly in parts of South America and Asia, i.e., in areas that do not correspond to the original concept of the geographical sites of gene centers (von Bothmer 1983).

The present state of the discussion is probably best interpreted by Harlan (1975). He classified crop species according to geographical patterns of variation as follows (cited from Lehmann 1984):

1. Endemic crop plants of a small, very restricted area. Centers of origin and of diversity coincide. Examples: Guinea millet (*Brachiaria deflexa* C.E. Hubb) and Black Fonio (*Digitaria iburua* Stapf) in Africa; Koda (*Paspalum scrobiculatum* L.) in India (Nilgiri mountains).

2. Semiendemic crop plants. The center of diversity sometimes shows either a faded center or nodal points. Examples: African rice (*Oryza glaberrima* Steud.), Teff (*Eragrostis tef* Trotter), Niger (*Guizotia abyssinica* (L.f) Cassini).
3. Monocentric crop plants of widespread distribution and a distinct center of origin, which can even be a center of diversity. Examples: coffee, rubber tree (*Hevea brasiliensis*), cacao, oil, palm.
4. Oligocentric crop plants with an extensive area. They are found in two or more centers of diversity. Examples: wheat, barley, peas, linseed, maize.
5. Noncentric crop plants distributed over a large area, but where no distinct center of origin or diversity can be pinpointed. Examples: sorghum, radish, rape, American *Phaseolus* beans, bottle gourd (*Lagenaria siceraria* (Molina) Standl.). The present state of knowledge is portrayed in Fig. 65.

It is clear then that centers of diversity are not static units, but are characterized by evolutionary processes that are continuing even in the present. As Harlan (1951) points out: "Evolution is proceeding at a rapid pace now" and "agronomists would be justified in accepting a more dynamic concept of the gene center." Kuckuck (1985) holds the following anthropogenic factors responsible for the dynamic population changes in these centers:

1. Seeds of populations are sometimes transported over long distances into areas of differing environments and varying conditions of natural selection. In the past, this was accomplished by caravans traveling, e.g., from India to the Bosporus, and nowadays by landowners via modern transportation.
2. Cultivation techniques are changed: in semiarid areas, wheat populations exhibit limited variation without irrigation; under irrigation, however, they evolve great diversity because of the lack of selection pressure.
3. Increased overgrazing also gradually changes the genetic composition of plant populations.
4. Growing related species site by site favors the repeated occurrence of spontaneous outcrossings, especially in regions of continental climates, and even in cleistogamous flowers of self-pollinating crops. If species hybrids derived from such crosses are more or less sterile, new fertile forms evolve from spontaneous backcrosses, mainly by substitution of single chromosomes or chromosome segments or by translocations (see Sect. 3.1.2).

The collector of plant germ plasm in a center of diversity is confronted with populations of various origin and, therefore, of different breeding value. Kuckuck (1974), for example, found the following wheat populations in Iran:

1. Indigenous populations exhibiting large variation.
2. Indigenous populations exhibiting small variation as a result of:
   (a) natural selection, e.g., by dry farming;
   (b) mass selection – origin of primitive landraces.

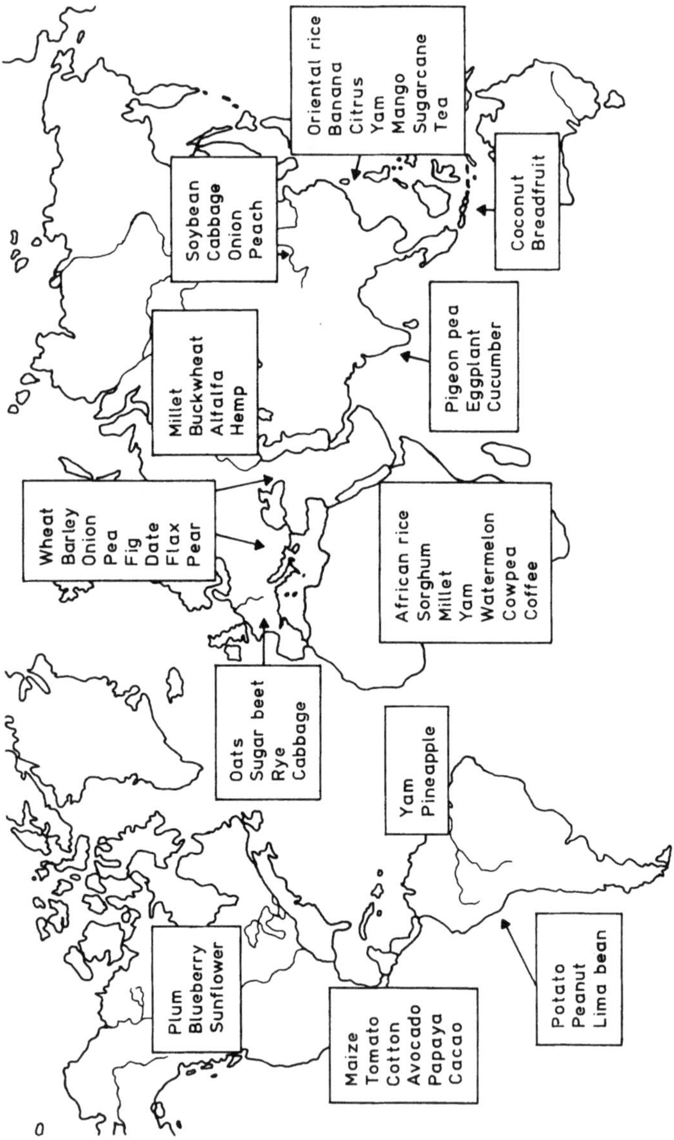

Fig. 65. Centers of genetic diversity (*shaded areas*) and the centers of origin of important crops (After Harlan 1986)

3. Uniform varieties, developed from indigenous populations by means of pedigree selection or from crosses of indigenous genotypes with introduced genotypes; sometimes these are no longer uniform as a result of mechanical contamination or spontaneous hybridization.
4. Uniform varieties introduced from foreign countries:
   (a) if introduced at least 35 years ago, they form a new starting point for genetic variation because of (mechanical) contamination by other genotypes, followed by spontaneous outcrossing, recombination, and mutation;
   (b) if introduced recently, they are of no value to the collector.

Thus, centers of diversity were seemingly unlimited resources of genetic variation, exploitable forever. Since the 1950s, however, action has been taken in the developing countries on a large scale to improve the standard of living by introducing modern techniques to agriculture. The activity in agriculture extended not only to improved soil management, the application of mineral fertilizers, and measures of plant protection, but also to the cultivation of modern cultivars.

In the beginning, these cultivars were introduced from developed countries. Later on, the developing countries began to grow their own cultivars and continued to do so with increasing frequency. Inevitably, the populations of landraces of these crop species vanished (see also Sect 4.5). Also, the constant pressure to bring even marginal land under cultivation has now started to threaten the continued existence of wild relatives and progenitors of any cultivated species. The danger of an irretrievable loss of genetic variation in the centers of diversity became obvious. Together with the narrowing of the genetic base caused by plant breeding itself, a severe crisis of further development in crop improvement was preordained to occur. This not only concerned general breeding activities, but especially the efforts initiated in the developing countries. Slogans like "genetics of disaster" or "genetic erosion" brought the situation at the time to a head.

6.4 Genetic Conservation

Though farsighted plant geneticists and breeders had recognized the value of plant genetic resources early on, measures undertaken to date have been solitary actions. Certainly in some countries notable collections were built up by interested institutes, but only in the USSR, due to the activities of Vavilov, and in the USA were coordinated measures taken to safeguard and utilize plant genetic resources. As a recent result of this, greater variability could be observed in barley in US collections than in those of 61 other countries.

Great credit must be given to Sir Otto Frankel; (CSIRO, Division of Plant Industry, Canberra) who, since the 1950s, has worked to devise effective measures against the loss of natural diversity via international cooperation (Frankel and Bennett 1970). Preliminary work was done by the Food and

Agricultural Organization of the United Nations (FA0) and by the International Biological Program (IBP).

To realize all the proposed measures, in 1974 the Consultative Group on International Agricultural Research (CGIAR), an international body promoting agricultural research in developing countries, founded the International Board for Plant Genetic Resources (IBPGR), which is located at FAO's headquarters in Rome. The Board received the mandate "to promote an international network of genetic resources centers to further collection, conservation, documentation, evaluation, and use of plant germ plasm . . . ". Guided by special advisory committees, the Board set priorities with respect to species and regions; emergency missions were made operational, technical advice for the establishment of storage facilities and documentation and data management was given, and training courses for personnel were supported. In the first 10 years, the establishment of plant genetic resources centers or units took precedence; since then, the evaluation of collected germ plasm has had priority. Worldwide, more than 100 institutions shall share the responsibility of a gene bank; about half of these are located in developing countries. Moreover, a network of base seed collections has been established for major crop species. Gene banks holding base collections act as either global or regional depositories of specific crops.

There are two basic principles involved in genetic or germ plasm conservation: (1) conservation in situ, and (2) conservation ex situ. Gene banks serve the function of ex situ conservation, in which seed accessions are stored under special conditions. Preferred storage conditions are a temperature of –18 °C and a seed moisture content of 5 ± 1% (wet basis) using air-tight containers. Under these conditions the viability of seeds will be preserved for many decades. All seeds of the orthodox type are adapted to storage; this type is characterized by a logarithmic growth of viability when seed moisture content and storage temperature are reduced. The vast majority of species of temperate climates belongs to this type. The recalcitrant type of seed, however, does not follow this principle: specifically the seeds cannot be dried without damage and they also display very limited natural viability; coffee and cacao are typical examples. Detailed information on the physiological and technological aspects of seeds as well as on gene bank management are given by Hanson (1985) and Ellis et al. (1985).

The conservation of germ plasm has the following functions: (1) as base collection for long-term conservation, and (2) as active collection for medium-term conservation, ca. 20 years, for direct use (delivery, evaluation, multiplication, etc.). Certainly in only a few cases does a spatial separation of the two functional units exist.

Stored seeds are kept under static conditions because natural evolution is interrupted. But the unchanged continuity for many years of once-described characters is attractive for plant breeding. On the other hand, the interruption of further evolutionary processes must be viewed critically if crop evolution is considered as a whole. Under these aspects and considering increasing genetic erosion, the establishment of gene banks can be seen as an emergency action. Another problem is the constantly increasing distance between stored and further evolving germ plasm in situ. Therefore, it is necessary to establish "renewable diversity," which solves the problem of the "stored diversity" of germ plasm and the "narrowed diversity" of breeding material.

Tests are being conducted, where possible under appropriate environmental conditions, to develop field gene banks, formerly called living collections, in various countries of the world. These gene banks must meanwhile serve vegetatively propagated species such as cassava, sweet potato, banana, citrus, cacao, sugarcane, *Allium*, and the perennial wild species of *Arachis* and *Glycine*, using modern agricultural practices. This finally will lead to population types that are adapted to intensive cultivation techniques.

Application of in vitro culture is especially suited for vegetatively propagated species and the recalcitrant seed storage type. At the present state of development, molecular techniques offer the possibility of conserving genes and genomes (DNA libraries). Furthermore, cells or tissues may be stored. From tissue cultures, whole plants can be regenerated in a number of cases. As has been pointed out in the chapter on variability in cell and tissue culture, in most cases, somaclonal variation occurs over the course of a long storage period. This phenomenon must be eliminated for in vitro cultures to preserve the genetic integrity of stored material.

For long-term storage, undifferentiated tissue (callus or suspensions, in some species young embryos as well) can be stored under extremely low temperatures using liquid nitrogen: this is called cryopreservation. Crystallization of cell sap during the freezing and dewing phase must be avoided; this can be accomplished with certain additives and by using precisely controlled temperature gradients. Even though the method is in use experimentally, it is not being applied routinely. This method would help to overcome the unwanted problem of variation.

The viability of stored seeds is determined by their genotypes. Even though today's handling of germ plasm in gene banks guarantees a long viability, routine monitoring is essential. If a multiplication of accessions becomes necessary, distinct genotypes or pure lines of self-pollinating species do not create any difficulties, unless they are adapted to certain environmental conditions such as photoperiodicity, cold tolerance, or other requirements that can be fulfilled only under special ecogeographic conditions. While formerly only those genotypes of a collected population were preserved that were of direct breeding value, in the meantime the preference is to store populations because which genotypes will be required for future programs cannot be foreseen.

The maintenance of the original composition of the gene pool of a self-pollinating species creates some problems if the multiplication cannot be performed in the country of origin. Under different environmental conditions, the gene pool or a population will be altered by natural selection. By changing gene frequencies, alleles can be fixed or eliminated, but losses within a self-pollinating species are irreparable. Easier is the situation of populations of cross-pollinating species. By recombination, existing genotypes can be recovered or even created anew. The only prerequisite is to avoid a biased selection by use of favorable environmental conditions. Further, a sufficient population size of self-pollinating as well as cross-pollinating species must be taken into account to avoid gene losses by genetic drift.

In the course, of handling and researching plant genetic resources, e.g., evaluation, a vast amount of data is available. These must be dealt with on a standardized basis to be compatible and usable, regardless of where. To manage this, standardized descriptors and corresponding descriptor states have been established. They are now used worldwide, as can be seen by a number of descriptor lists for various crop species.

6.5 Cited and Recommended Literature for Chapter 6

Bosemark NO (1979) Genetic poverty of the sugarbeet in Europe. In: Proc Conf on Broadening the Genetic Base of Crops. PUDOC, Wageningen, pp 29–35

Bothmer Rv (1983) Germplasm collections and status of wild species of *Hordeum* (barley). Sv Utsädesföreningens Tidskr 93:145–150

Brieger F (1963) Collections and evaluations of indigenous races of maize. Genet Agrar 17:259–264

Brücher H (1969) Gibt es Genzentren? Naturwissenschaften 56:77–84

Cox TS, Murphy JP, Rodgers DM (1986) Changes in genetic diversity in the red winter regions of the United States. Proc Natl Acad Sci USA 83:5583–5586

Duvick DN (1984) Genetic diversity in major farm crops on the farm and in reserve. Econ Bot 38:161–178

Ellis RH, Hong TD, Roberts EH (1985) Handbook of seed technology for genebanks (Handbook for genebanks No 2), vol I. Principles and methodology, vol II. Compendium of specific germination information and test recommendations. IBPGR, Rome

Frankel OH, Bennet E (eds) (1970) Genetic resources in plants – their exploration and conservation (IBP Handbook No 11) Blackwell Sci Publ, Oxford

Frankel OH, Hawkes JG (eds) (1975) Crop genetic resources for today and tomorrow. CIBP No 2, Cambridge Univ Press, Cambridge

Hanson J (1985) Procedure for handling seeds in genebanks (Practical manual for genebanks No 1). IBPGR, Rome

Harlan JR (1951) Anatomy of gene centers. Am Nat 85:97–103

Harlan JR (1975) Crops and man. Am Soc Agronomy, Madison

Harlan JR (1986) The plants and animals that nourish man. Sci Am 235:88–97

International Board for Plant Genetic Resources (IBPGR) (1975–1988) Plant Genetic Resources Newslett, Rome

Kuckuck H (1974) Bedeutung, Erhaltung und Weiterentwicklung der genetischen Formenmannigfaltigkeit für die Pflanzenzüchtung – ein Beitrag zür Grünen Revolution. Naturwiss Rundsch 27:267–272

Kuckuck H (1985) Erhaltung und Nutzung der natürlichen Formenmannigfaltigkeit. In: Kuckuck H, Kobabe G, Wenzel G (1985) Grundzüge der Pflanzenzüchtung, 5. Aufl. de Gruyter Berlin, pp 211–219

Lehmann CO (1984) Die Genzentrentheorie in heutiger Sicht. In: Aktuelle Probleme der landwirtschaftlichen Forschung, 9. Seminar (1982), Abstammung der Kulturpflanzen und die Erhaltung des natür lichen Formenreichtums. Landw-Chem Bundesanst Linz, pp 28–36

Maly R, Hammer K, Lehmann CO (1987) Sammlung pflanzlicher genetischer Ressourcen in Süditalien – ein Reisebericht aus dem Jahr 1950 mit Bemerkungen zum Schicksal der Landsorten "in situ" und in der Genbank. Kulturpflanze 35:109–134

Pierce RO, Jain SK (1977) Variation in some plant and seedoil characteristics of meadowfoam. Crop Sci 17:521–526

Ross H (1986) Potato breeding – problems and perspectives. Adv Plant Breed Suppl 13 Parey, Berlin

Schwarzbach E (1986) Verwendung von Wildgerstenmaterial in der Züchtung. In. Verfügbarkeit genetischer Resourcen, AG Pflanzenzüchtung, Vortr Pflanzenzücht 10:60–64

Sneep J, Hendriksen AJT (eds) (1979) Plant breeding perspectives. Centre for Agric Publ and Documentation, Wageningen

Vavilov NI (1928) Geographische Zentren unserer Kulturpflanzen. Z Ind Abst Vererbungsl, Suppl 1:342–369

Subject Index

MIX
Papier aus verantwortungsvollen Quellen
Paper from responsible sources
FSC® C105338

If you have any concerns about our products,
you can contact us on
ProductSafety@springernature.com

In case Publisher is established outside the EU,
the EU authorized representative is:
Springer Nature Customer Service Center GmbH
Europaplatz 3, 69115 Heidelberg, Germany

Printed by Libri Plureos GmbH
in Hamburg, Germany